MATHEMATICAL LOGIC AND FOUNDATIONS OF SET THEORY

STUDIES IN LOGIC

AND

THE FOUNDATIONS OF MATHEMATICS

Editors

A. HEYTING, *Amsterdam*
A. MOSTOWSKI, *Warszawa*
A. ROBINSON, *New Haven*
P. SUPPES, *Stanford*

Advisory Editorial Board

Y. BAR-HILLEL, *Jerusalem*
K. L. DE BOUVÈRE, *Santa Clara*
H. HERMES, *Freiburg i/Br.*
J. HINTIKKA, *Helsinki*
J. C. SHEPHERDSON, *Bristol*
E. P. SPECKER, *Zürich*

NORTH-HOLLAND PUBLISHING COMPANY
AMSTERDAM • LONDON

MATHEMATICAL LOGIC AND FOUNDATIONS OF SET THEORY

PROCEEDINGS OF AN INTERNATIONAL COLLOQUIUM
HELD UNDER THE AUSPICES OF
THE ISRAEL ACADEMY OF SCIENCES AND HUMANITIES

JERUSALEM, 11–14 NOVEMBER 1968

Edited by
YEHOSHUA BAR-HILLEL
Professor of Logic and Philosophy of Science
The Hebrew University of Jerusalem, Israel

1970

NORTH-HOLLAND PUBLISHING COMPANY
AMSTERDAM • LONDON

Library of Congress Catalog Card Number 73–97195

ISBN 7204 2255 8

PUBLISHERS:

NORTH-HOLLAND PUBLISHING COMPANY — AMSTERDAM

NORTH-HOLLAND PUBLISHING COMPANY, LTD — LONDON

PRINTED IN ISRAEL

PREFACE

This volume comprises seven of the eight addresses presented before the International Colloquium on Mathematical Logic and Foundations of Set Theory held at the Academy Building in Jerusalem, Israel, on November 11–14, 1968. The Colloquium was sponsored by the Israel Academy of Sciences and Humanities, the Mathematical Institute of the Hebrew University of Jerusalem, and the International Mathematical Union, and was dedicated to the memory of Professor Abraham A. Fraenkel, one of the founders of axiomatic set theory, the beloved teacher of three of the eight invited speakers and of the editor, and a founding member of the Israel Academy.

The bulk of the support for the Colloquium was granted by the International Mathematical Union, while additional aid came from the Israel Academy which also served as the host of the Colloquium and whose staff was most helpful with regard to all technical arrangements.

At the opening public session of the Colloquium, held on the evening of November 11, at the Hebrew University of Jerusalem, Professor Y. Bar-Hillel presented (in Hebrew) an appreciation of Professor Fraenkel's contribution to the Philosophy of Mathematics, and Professor Alfred Tarski gave a lecture on Some Reflections on Recent Developments in the Foundations of Set Theory.

No attempt was made to unify the contributors' notation, terminology and bibliographical style, but it is the editor's feeling that no appreciable harm was caused thereby; on the other hand, it facilitated the speedy publication of this volume. Similarly, it was quickly decided that no index was called for.

CONTENTS

WEAKLY DEFINABLE RELATIONS AND SPECIAL AUTOMATA*

BY

MICHAEL O. RABIN

In this paper we consider monadic second-order theories and study problems of definability. As a by-product we obtain certain decidability results. Let $\mathcal{N}_2 = \langle T, r_0, r_1 \rangle$ be the structure of two successor functions (see §1). Let L be the monadic second-order language appropriate for $\mathcal{N}_2$ which has individual variables $\boldsymbol{x}, \boldsymbol{y}, \boldsymbol{z}, \ldots$, ranging over elements of T, finite-set variables $\boldsymbol{\alpha}, \boldsymbol{\beta}, \boldsymbol{\gamma}, \ldots$, ranging over finite subsets of T, and set variables $\boldsymbol{A}, \boldsymbol{B}, \boldsymbol{C}, \ldots$, ranging over arbitrary subsets of T. A relation $H \subseteq P(T)^n$ between subsets of T is called *definable* in the second-order theory (language) of $\mathcal{N}_2$ if for some formula $F(\boldsymbol{A}_1, \ldots, \boldsymbol{A}_n)$ of L

$$(1) \qquad H = \{(A_1, \ldots, A_n) \,|\, (A_1, \ldots, A_n) \in P(T)^n, \mathcal{N}_2 \models F(A_1, \ldots, A_n)\}.$$

The relation H is *weakly-definable* if (1) holds for a formula $F(\boldsymbol{A}_1, \ldots, \boldsymbol{A}_n)$ containing just individual and finite-set quantifiers.

In [6] we have characterized the definable relations by means of finite automata operating on infinite trees. This result was used to solve the decision problem of the second-order theory of $\mathcal{N}_2$. This in turn entailed the decidability of many theories.

Here we introduce the notion of a *special automaton* on infinite trees and use it to characterize the weakly definable sets. An automaton on infinite trees may be viewed as representing a relation $H \subseteq P(T)^n$ for some n. It turns out that a relation $H \subseteq P(T)^n$ is weakly definable if and only if both H and complement $P(T)^n - H$ are represented by appropriate special automata. On the other hand, not every relation H represented by a special automaton is weakly definable. Rather, a relation $H \subseteq P(T)^n$ is represented by a special automaton if and only if (1) holds with a formula $F(\boldsymbol{A}_1, \ldots, \boldsymbol{A}_n)$ in prenex form which has only existential arbitrary-set quantifiers. This yields the following syntactical result. A formula $F(\boldsymbol{A}_1, \ldots, \boldsymbol{A}_n)$ is equivalent (in $\mathcal{N}_2$) with some formula $G(\boldsymbol{A}_1, \ldots, \boldsymbol{A}_n)$ containing only finite-set quantifiers, if and only if F is equivalent to some prenex formula $F_1(\boldsymbol{A}_1, \ldots, \boldsymbol{A}_n)$ having only existential arbitrary-set quantifiers, and also to some prenex formula $F_2(\boldsymbol{A}_1, \ldots, \boldsymbol{A}_n)$ having only universal arbitrary-set quantifiers.

* This research was sponsored under Contract No. N00014 69 C 0192, U.S. Office of Naval Research, Information Systems Branch, in Jerusalem.

As a by-product of the characterization of weakly definbale relations we get the solution of certain decision problems. In [6] we have shown that the weak second-order theory of a unary function, and the weak second-order theory of linearly ordered sets (see [4]), arc decidable. These results were actually corollaries of stronger theorems concerning the corresponding full monadic second-order theories. Here we deduce the same decidability results using the information concerning weakly definable relations and special automata. Also the many applications by D. M. Gabbay of [3] to the solution of the decision problem of various logical calculi follow already from Theorem 24 of the present paper.

1. Notations and basic standard definitions

We shall employ the standard notations and terminology concerning sets, mappings, structures, and logical calculi, used in [6].

As usual, each natural number n is the set of all smaller numbers. Thus $0 = \emptyset$, $1 = \{0\}$, $2 = \{0,1\}$, and $n = \{0,1,\ldots,n-1\}$. An *n-termed sequence* is a mapping $x\colon n \to A$. The sequence x is also called a *word* on A. The ith coordinate of the sequence is $x(i)$, $0 \leqq i < n$, and will sometimes be denoted by x_i. The *length* $l(x)$ of x is $l(x) = n$. The sequence x will also be written as $(x_0,\ldots,x_{n-1})$. If $x = (x_0,\ldots,x_{n-1})$ and $y = (y_0,\ldots,y_{m-1})$ then xy will denote the sequence $(x_0,\ldots,x_{n-1}, y_0,\ldots,y_{m-1})$. We have $l(xy) = l(x) + l(y)$. The sequence (x_0) of length one will also be written as x_0. Thus $x = x_0 x_1 \ldots x_{n-1}$. The unique empty sequence of length 0 will be denoted by Λ.

For each $i < \omega$, the *projection* p_i is the function which is defined by $p_i(x) = x_i$ for $x = (x_0,\ldots,x_{n-1})$, $i < n$.

The *infinite binary tree* is the set $T = \{0,1\}^*$ of all finite words on $\{0,1\}$. The elements $x \in T$ are the *nodes* of T. For $x \in T$, the nodes $x0, x1$ are called the *immediate successors* of x. The empty word Λ is called the *root* of T. Our language is suggested by the following picture. The lowest node of T is the root Λ. The root branches up to the (say) left into the node 0 and to the right into the node 1. The node 0 branches into 00 and 01; the node 1 branches into 10 and 11. And so on ad infinitum.

On T we define a partial-ordering by $x \leqq y$ (x is an *initial* of y) if and only if $\exists z[y = xz]$. If $x \leqq y$ and $x \neq y$ then we shall write $x < y$.

For $x \in T$, the *subtree* T_x with *root* x is defined by $T_x = \{y \mid y \in T, x \leqq y\}$. Thus $T_\Lambda = T$.

A *path* π of a tree T_x is a set $\pi \subset T_x$ satisfying 1) $x \in \pi$; 2) for $y \in \pi$, either $y0 \in \pi$ or $y1 \in \pi$, but not both; 3) π is the smallest subset of T_x satisfying 1–2.

Note that if $\pi \subset T$ is a path and $x, y \in \pi$, then $x \leqq y$ or $y \leqq x$.

A subset $F \subset T_x$ is called a *frontier* of T_x if for every path $\pi \subset T_x$ we have $c(\pi \cap F) = 1$. It is readily seen that if $F \subset T_x$ is a frontier then F is finite. If $F_1 \subset T_x$ and $F_2 \subset T_x$ are frontiers we shall say that F_2 *is bigger than* F_1 ($F_1 < F_2$) if for every $y \in F_2$ there exists a $x \in F_1$ such that $x < y$. For $S \subseteq T$ we have $c(S \cap \pi) = \omega$ for every path $\pi \subset T$, if and only if $S = \bigcup_{n<\omega} F_n$ where F_n is a frontier of T and $F_n < F_{n+1}$, $n < \omega$.

A *finite (frontiered) tree* is a set $E = \{x \mid x \leqq y \text{ for some } y \in F\}$ where F is a fixed frontier of T. For E as above, F is called the *frontier* of E and denoted by $Ft(E)$. By "finite tree" we shall always mean a finite frontiered tree.

For $a \in \{0,1\}$ define the (immediate) *successor function* $r_a: T \to T$ by $r_a(x) = xa$, $x \in T$. The *structure of two successor functions* is $\mathcal{N}_2 = \langle T, r_0, r_1 \rangle$.

With $\mathcal{N}_2$ we associate an appropriate (monadic) second-order language L_2. This L_2 has function-constants $\boldsymbol{r}_0, \boldsymbol{r}_1$, to denote r_0 and r_1; the usual logical connectives and quantifiers; the membership symbol $\in$; equality; individual variables $\boldsymbol{x}, \boldsymbol{y}, \boldsymbol{z}, \ldots$, ranging over elements of T; finite-set variables $\boldsymbol{\alpha}, \boldsymbol{\beta}, \boldsymbol{\gamma}, \ldots$, ranging over finite subsets of T; set variables $\boldsymbol{A}, \boldsymbol{B}, \boldsymbol{C}, \ldots$, ranging over arbitrary subsets of T. The atomic formulas of L_2 include formulas of the form $t \in V$ where t is a term of L_2 and V is a (finite or arbitrary) set variable. Quantification is possible over all the three sorts of variables.

The *second-order theory of two successor functions* (S2S) is the set of all sentences F of L_2 such that $\mathcal{N}_2 \models F$ (F is true in $\mathcal{N}_2$). The theory S2S was proved decidable in [6] by means of a theory of automata on infinite trees.

DEFINITION 1. An n-ary relation $R \subseteq P(T)^n$ between subsets of T is *definable* in L_2 (S2S) if there exists a formula $F(\boldsymbol{A}_1, \ldots, \boldsymbol{A}_n)$ of L_2 such that

$$R = \{(A_1, \ldots, A_n) \mid \mathcal{N}_2 \models F(A_1, \ldots, A_n)\}. \tag{1}$$

The relation R is *weakly-definable* if (1) holds for a formula containing quantifiers only over individual and finite-set variables.

2. Special automata

As stated in the Introduction, our aim is to characterize the weakly-defined relations. To this end we develop a theory of special automata.

In the following, Σ denotes a finite set called the *alphabet*.

DEFINITION 2. A Σ-*(valued)tree* is a pair (v, T_x) such that $v: T_x \to \Sigma$. If (v, T) is a valued tree then (v, T_x) will denote the induced valued subtree $(v \mid T_x, T_x)$. The set of all Σ-trees (v, T_x), for a fixed $x \in T$, will be denoted by $V_{\Sigma,x}$. The set $\bigcup_{x \in T} V_{\Sigma,x}$ of all Σ-trees will be denoted by V_Σ.

DEFINITION 3. A *table* over Σ-trees is a pair $\langle S, M\rangle$ where S is a finite set, the *set of states*, and M is a function $M: S \times \Sigma \to P(S \times S)$, the (non-deterministic) *table of moves* ($P(A)$ denotes the set of all subsets of A).

A *special finite automaton* (s.f.a.) over Σ-trees (a *special Σ-automaton*) is a system $\mathfrak{A} = \langle S, M, S_0, F\rangle$ where $\langle S, M\rangle$ is as above, $S_0 \subseteq S$ is the set of *initial states*, $F \subseteq S$ is the set of *designated states*.

DEFINITION 4. A *run* of $\langle S, M\rangle$ on the Σ-tree $t = (v, T_x)$ is a mapping $r: T_x \to S$ such that for $y \in T_x$, $(r(y0), r(y1)) \in M(r(y), v(y))$. We also talk about a run of an automaton $\mathfrak{A}$ on a tree, meaning a run of the associated table. The set of all $\mathfrak{A}$-runs on t is denoted by $\mathrm{Rn}(\mathfrak{A}, t)$.

For a mapping $\phi: A \to B$ define $\mathrm{In}(\phi) = \{b \mid b \in B,\ c(\phi^{-1}(b)) \geqq \omega\}$.

DEFINITION 5. The special automaton $\mathfrak{A} = \langle S, M, S_0, F\rangle$ *accepts* (v, T_x) if there exists an *accepting* $\mathfrak{A}$-run r on (v, T_x) such that $r(x) \in S_0$ and for every path π of T_x, $\mathrm{In}(r \mid \pi) \cap F \neq \emptyset$. The set $T(\mathfrak{A})$ of Σ-trees *defined* by $\mathfrak{A}$ is

$$T(\mathfrak{A}) = \{(v, T_x) \mid x \in T,\ (v, T_x) \textit{ accepted by } \mathfrak{A}\}.$$

A set $A \subseteq V_\Sigma$ is *s.f.a. definable* if for some s.f.a. $\mathfrak{A}$, $A = T(\mathfrak{A})$.

REMARK. It is quite clear that the special automata introduced here are a weaker version of the automata defined in [6]. For every special automaton $\mathfrak{A}$ there exists an automaton $\mathfrak{B}$ in the sense of [6], such that $T(\mathfrak{A}) = T(\mathfrak{B})$. In fact, $\mathfrak{B}$ may be taken to have the same table and initial status as $\mathfrak{A}$. That the converse statement is not true is shown by the example in §3.

REMARK. A set $A \subset V_\Sigma$ is called *invariant* if for every Σ-tree $t = (v, T)$ and every $x \in T$, $t \in A$ if and only if the tree $t' = (v', T_x)$ defined by $v'(xy) = v(y)$, $y \in T$, is in A. The invariant subsets of V_Σ are a boolean algebra. It is clear from Definition 5 that every set $T(\mathfrak{A})$ is invariant. To prove that an invariant set A is s.f.a. definable, it suffices to construct an automaton $\mathfrak{A}$ such that $(v, T) \in T(\mathfrak{A})$ if and only if $(v, T) \in A$.

The following results are immediate.

LEMMA 1. *If $A \subseteq V_\Sigma$ is s.f.a. definable, then there exists an automaton $\mathfrak{A} = \langle S, M, S_0, F\rangle$ such that $S_0 = \{s_0\}$, $s_0 \in S$, and $T(\mathfrak{A}) = A$. This $\mathfrak{A}$ may be chosen so that $(s_1, s_2) \in M(s, \sigma)$ implies $s_1 \neq s_0$, and $s_2 \neq s_0$.*

THEOREM 2. *If $A, B \subseteq V_\Sigma$ are s.f.a. definable, then so are $A \cup B$ and $A \cap B$.*

Proof. Let $A = T(\mathfrak{A})$, $B = T(\mathfrak{B})$ where $\mathfrak{A} = \langle S, M, s_0, F\rangle$, $\mathfrak{B} = \langle S', M', s'_0, F'\rangle$; we assume that $S \cap S' = \emptyset$. Construct the automaton

$$\mathfrak{A} \cup \mathfrak{B} = \langle S \cup S',\ M \cup M',\ \{s_0, s_0'\},\ F \cup F'\rangle.$$

Clearly, $T(\mathfrak{A} \cup \mathfrak{B}) = A \cup B$.

With the above notations, define $\mathfrak{A} \times \mathfrak{B} = \langle S \times S' \times \{0,1,2\}, \bar{M}, (s_0, s_0', 0), F\rangle$ as follows. $((s_1, s_1', b), (s_2, s_2', b)) \in \bar{M}((s, s', a), \sigma)$ if and only if $(s_1, s_2) \in M(s, \sigma)$, $(s_1', s_2') \in M'(s', \sigma)$; $b = 1$ if and only if $a = 0$ and $s \in F$, or $a = 1$ and $s' \notin F'$; $b = 2$ if and only if $a = 1$ and $s' \in F'$; $b = 0$ if and only if $a = 2$, or $a = 0$ and $s \notin F$. Put $\bar{F} = S \times S' \times \{2\}$. We have $T(\mathfrak{A} \times \mathfrak{B}) = A \cap B$.

Definition 6. Let $t = (v, T)$ be a $\Sigma_1 \times \Sigma_2$-tree and let p_0 be the projection $p_0(x, y) = x$. The *projection* $p_0(t)$, by definition, is the Σ_1-tree $(p_0 v, T)$.

The *projection* $p_0(A)$ of a set $A \subseteq V_{\Sigma_1 \times \Sigma_2}$, is $p_0(A) = \{p_0(t) \mid t \in A\}$. The Σ_2-*cylindrification* of a set $B \subseteq V_{\Sigma_1}$ is the largest set $A \subset V_{\Sigma_1 \times \Sigma_2}$ such that $p_0(A) = B$.

Theorem 3. *If $A \subseteq V_{\Sigma_1 \times \Sigma_2}$ is a s.f.a. definable set, then $p_0(A) \subseteq V_{\Sigma_1}$ is a s.f.a. definable set. If $B \subseteq V_{\Sigma_1}$ is s.f.a. definable, so is its Σ_2-cylindrification $A \subseteq V_{\Sigma_1 \times \Sigma_2}$.*

Proof. Let $\mathfrak{A} = \langle S, M, s_0, F\rangle$ be a $\Sigma_1 \times \Sigma_2$-automaton with $T(\mathfrak{A}) = A$. Define a Σ_1-automaton by $\mathfrak{A}_1 = \langle S, M_1, s_0, F\rangle$, where

$$M_1(s, \sigma_1) = \bigcup_{\sigma_2 \in \Sigma_2} M(s, (\sigma_1, \sigma_2)), \quad \sigma_1 \in \Sigma_1,\ s \in S.$$

One can check that $T(\mathfrak{A}_1) = p_0(A)$.

The proof concerning cylindrification is left to the reader.

3. A counterexample

We wish to show that the class of s.f.a. definable sets is not closed under complementation. We shall exhibit a s.f.a. definable set $B \subseteq V_\Sigma$ such that $A = V_\Sigma - B$ is not s.f.a. definable.

Let $\Sigma = \{0, 1\}$ and let B be the set of all Σ-trees (v, T_x) such that for some path $\pi \subset T_x$ we have: $1 \in \mathrm{In}(v \mid \pi)$. It is readily seen that B is s.f.a. definable.

The set $A = V_\Sigma - B$ consists of all Σ-trees (v, T_x) such that for every path $\pi \subset T_x$, $1 \notin \mathrm{In}(v \mid \pi)$. We claim that A is not s.f.a. definable. To prove this, let us assume that $\mathfrak{A} = \langle S, M, s_0, F\rangle$ is a special automaton such that $T(\mathfrak{A}) = A$ and derive a contradiction Throughout this section, unless otherwise specified, $\mathfrak{A}$ will denote this particular automaton. We shall need the following construction.

DEFINITION 7. Let $t = (v, T)$ and $t_1 = (v_1, T_x)$, $x \in T$, be Σ'-trees. The result of *grafting* the tree t_1 on t at $y \in T$ is the tree (v_2, T) such that $v_2(z) = v(z)$ for $x \notin T_y$, and $v_2(yz) = v_1(xz)$ for $z \in T$. (Note that $T_y = \{yz \mid z \in T\}$, and similarly for T_x.)

DEFINITION 8. Let $t_n = (v_n, T)$, $n < \omega$, be Σ'-trees. We shall say that $\lim_{n\to\infty} t_n = (v, T)$ if there exists an integral valued function $N(x)$, $x \in T$, such that $N(x) \leqq n$ implies $v_n(x) = v(x)$.

LEMMA 4. *Let $t = (v, T) \in T(\mathfrak{A})$ and let $r \in \mathrm{Rn}(\mathfrak{A}, t)$ be an accepting run If there exist nodes $x < z < y$ such that $r(x) = r(y) = s$, $s \in F$, and $v(z) = 1$, then there exists a tree $t' \notin A$ which is accepted by $\mathfrak{A}$.*

Proof. Assume that $y = xu$. Let $\Sigma' = S \times \Sigma$, and let (ϕ, T) be the Σ'-tree such that $\phi(z) = (r(z), v(z))$, $z \in T$. Graft $t_x = (\phi, T_x)$ on (ϕ, T) at the node $y = xu$ and call the resulting tree $t_1 = (\phi_1, T)$. Since $r(x) = r(y)$, we have that $p_0\phi_1$ is an $\mathfrak{A}$-run on $(p_1\phi_1, T)$.

Note that $p_0\phi_1(xu^2) = s$. Graft t_x on t_1 at xu^2 to obtain $t_2 = (\phi_2, T)$. Again $p_0\phi_2$ is an $\mathfrak{A}$-run on $(p_1\phi_2, T)$ and $p_0\phi_2(xu^3) = s$. Continue this process inductively for every $n < \omega$, where at the nth step we graft t_x on t_{n-1} at xu^n to obtain $t_n = (\phi_n, T)$. Let $\lim_{n\to\infty} t_n = \bar{t} = (\bar{\phi}, T)$, and $t' = (p_1\bar{\phi}, T)$. Since each $p_0\phi_n$ is an $\mathfrak{A}$-run on $(p_1\phi_n, T)$, $\bar{r} = p_0\bar{\phi}$ is an $\mathfrak{A}$-run on t'.

We claim that for every path $\pi \subset T$, $\mathrm{In}(\bar{r} \mid \pi) \cap F \neq \emptyset$ so that $t' \in T(\mathfrak{A})$. Namely, if $xu^n \in \pi$ for infinitely many (and hence all) $n < \omega$ then $\bar{r}(xu^n) = s$ and $s \in \mathrm{In}(\bar{r} \mid \pi) \cap F$. Otherwise, two cases may occur. Either π contains no xu^n, then $\bar{r} \mid \pi = r \mid \pi$ and $\mathrm{In}(\bar{r} \mid \pi) \cap F = \mathrm{In}(r \mid \pi) \cap F \neq \emptyset$. Or else there is an $n < \omega$ such that $xu^n \in \pi$, $xu^{n+1} \notin \pi$. In this case there exists a path $\pi' \subset T_x$ such that $\bar{r}(xu^n v) = r(xv)$ for all $xv \in \pi$. Thus $\mathrm{In}(\bar{r} \mid \pi) \cap F = \mathrm{In}(r \mid \pi') \cap F \neq \emptyset$.

Now, if $z = xw$ then $p_1\bar{\phi}(xu^n w) = 1$, $n < \omega$; hence $t' \notin A$.

Let $t = (v, T)$ be a $\{0,1\}$-tree. We define, by induction on n, subsets $D_n(t) \subseteq T$. $D_0(t) = \{x \mid v(x) = 1\}$; $D_{n+1}(t) = D_n(t) \cap \{x \mid c(D_n(t) \cap T_x) = \omega\}$.

LEMMA 5. *Let $\mathfrak{A} = \langle S, M, s_0, F \rangle$ be a special $\{0,1\}$-automaton; $t = (v, T) \in V_{\{0,1\}}$, $\Lambda \in D_{n+1}(t)$. If $r \in \mathrm{Rn}(\mathfrak{A}, t)$ is an accepting run such that $c(r(T - \{\Lambda\})) \leqq n$, then for some $x < z < y$, $r(x) = r(y) = s$, $s \in F$, and $r(z) = 1$.*

Proof. By induction on n. For every path π, let $x(\pi)$ be the first $\Lambda < x \in \pi$ such that $r(x) \in F$. The set $\{x(\pi) \mid \pi \subset T\}$ is a frontier and hence $E = \{y \mid y \leqq x(\pi)$, *for some* $\pi \subset T\}$, is finite. Since $c(D_n(t) \cap T) = \omega$, there must be a point $z \in D_n(t)$ such that $z \notin E$ and hence for some $x = x(\pi)$, $x < z$. Now $r(x) = s$, $s \in F$ and $v(z) = 1$. If for some $z < y$, $r(y) = s$.

then we are through. Otherwise $s \notin r(T_z - \{z\})$ and, since $s \in r(T - \{\Lambda\})$, $c(r(T_z - \{z\})) \leqq n-1$. Since $z \in D_n(t)$, and $r \mid T_z$ is an accepting run of $\langle S, M, r(z), F \rangle$ on (v, T_z), the proof is completed by induction.

LEMMA 6. *For every* $1 \leqq n$ *there exists a tree* $t_n \in A$ *such that* $\Lambda \in D_{n-1}(t_n)$.

Proof. Let $t_1 = (v_1, T)$, $v_1(\Lambda) = 1$, $v_1(x) = 0$, $\Lambda < x$. The tree t_2 is obtained by grafting t_1 on t_1 at each of the nodes $1^k 0$, $1 \leqq k < \omega$. In general, t_n is obtained by grafting t_{n-1} on t_1 at each of the above mentioned nodes.

LEMMA 7. *The set* A *is not s.f.a.-definable.*

Proof. Assume $A = T(\mathfrak{A})$ for $\mathfrak{A} = \langle S, M, s_0, F \rangle$ where $c(S) = n$. Then t_{n+2} is accepted by $\mathfrak{A}$. By Lemma 5, there exist nodes $x < z < y$ satisfying the conditions of Lemma 4. Hence there exists a $t' \notin A$ which is accepted by $\mathfrak{A}$, a contradiction.

COROLLARY 8. *Not every set of valued trees definable by an automaton (in the sense of* [6]) *is definable by a special automaton. The set* A *is an example for this.*

4. A closure result

THEOREM 9. *Let* $\mathfrak{A} = \langle S, M, s_0, F \rangle$ *be a special* Σ*-automaton, and let* $\tau \in \Sigma$. *Define* D *to be the set of all* Σ*-trees* $t = (v, T)$ *such that* $t \in D$ *if and only if for every* $x \in T$, $v(x) = \tau$ *implies* $(v, T_x) \in T(\mathfrak{A})$. *The set* D *is s.f.a. definable.*

We need some definitions and lemmata to prove this result. The intuitive approach to the task of recognizing whether $t \in D$, is to start a copy of $\mathfrak{A}$ at each node x such that $v(x) = \tau$, and check whether (v, T_x) is accepted by $\mathfrak{A}$.

This process, however, cannot be accomplished by a finite automaton.

The crucial observation is that for any $y \in T$, even though many copies of an $\mathfrak{A}$ may have been activated at various $x < y$, at y the number of *different* states of $\mathfrak{A}$ which appear is still bounded by the cardinality of the set S. Thus, all the copies of $\mathfrak{A}$ reaching y in the same state s can be replaced by just one of these copies. In this way, we have, at any node y, just a bounded number of copies of $\mathfrak{A}$, and this can be described by a finite Σ-table. In addition to having copies of $\mathfrak{A}$ move on (v, T), we will also need to record which copies merged when reaching the same state. The above considerations motivate the following formal definition of a Σ-table.

Let $u \in S^*$ be a finite sequence on S. For $i < m = l(u)$, put $u(i)' = u(i)$ if $u(i) \neq u(j)$ for $j < i$, $u(i)' = \Lambda$ otherwise. Define $C(u)$, *the contraction of* u, by $C(u) = u(0)'u(1)' \ldots u(m-1)'$.

Define the *contracting* function $\phi_u\colon l(u) \to l(C(u))$ by $\phi_u(i) = j$ if and only if $u(i) = C(u)(j)$. Clearly ϕ_u maps $l(u)$ onto $l(C(u))$. Furthermore, $\phi_u(i) \leqq i$ for $i < l(u)$; and if $\phi_u(i) < i$ then $\phi_u(j) < j$ for $i \leqq j < l(u)$.

A word $u \in S^*$ is called *contracted* if $i < j < l(u)$ implies $u(i) \neq u(j)$. The word u is contracted if and only if $C(u) = u$. The word $C(u)$ is contracted for all $u \in S^*$. If u is contracted then $l(u) \leqq c(S)$.

Let $\mathfrak{A} = \langle S, M, s_0, F \rangle$ be a special Σ-automaton. Assume $c(S) = n$. Without loss of generality, we assume that $\mathfrak{A}$ is such that $(s_1, s_2) \in M(s, \sigma)$ implies $s_1 \neq s_0$, $s_2 \neq s_0$ (see Lemma 1).

If $f\colon A \to B$ then $D(f)$ and $R(f)$ will denote, respectively, the domain A and range $f(A)$ of f.

DEFINITION 9. For $\mathfrak{A}$ and $\tau \in \Sigma$ as above, define the Σ-table $\mathfrak{B} = \langle S^{\mathfrak{B}}, M^{\mathfrak{B}} \rangle$ as follows:

Set $U = \{u \mid u \in S^*,\ \ C(u) = u,\ \ u(i) = s_0 \ \ \textit{implies}\ \ i = l(u) - 1\}$; $H = \{\phi \mid \phi\colon m \to k,\ R(\phi) = k,\ k \leqq m < n\}$. Define $S^{\mathfrak{B}} = (U \times H) \cup \{d\}$, where d is a *dump state*.

Define $M^{\mathfrak{B}}$ by cases. $M^{\mathfrak{B}}(d, \sigma) = \{(d,d)\}$ for $\sigma \in \Sigma$. Let $(u, \phi) \in S^{\mathfrak{B}}$, $l(u) = k$, and $\sigma \in \Sigma$. $M^{\mathfrak{B}}(u, \phi), \sigma) = \{(d,d)\}$ if $u(k-1) = s_0$ and $\sigma \neq \tau$, or $u(k-1) \neq s_0$ and $\sigma = \tau$.

In all other cases, $((u_1, \phi_1), (u_2, \phi_2)) \in M^{\mathfrak{B}}((u, \phi), \sigma)$ if and only if for some $w_1, w_2 \in S^k$ we have $(w_1(i), w_2(i)) \in M(u(i), \sigma)$ for $i < k$; $u_j = C(w_j)a_i$ where $a_j \in \{\Lambda, s_0\}$, $j = 1,2$; $\phi_j = \phi_{w_j}$, $j = 1,2$.

REMARK. It is readily seen that, with the above notations,

$$(u_1(\phi_1(i)),\ u_2(\phi_2(i))) \in M(u(i), \sigma),\ \ i < l(u).$$

The idea behind this definition is as follows. There are up to n copies of $\mathfrak{A}$ scanning each node x of T. If $(u, \phi) \in S^{\mathfrak{B}}$, is the state at $x \in T$, $l(u) = k$, then copies $0, 1, \ldots, k-1$ are active at x and in states $u(0), u(1), \ldots, u(k-1)$. If $v(x) = \sigma$ then, unless $u(k) = s_0$ if and only if $\sigma = \tau$, $\mathfrak{B}$ moves into a dump state d, i.e. $M^{\mathfrak{B}}((u, \phi), \sigma) = \{(d,d)\}$. Thus for $\mathfrak{B}$ to avoid the state d, a copy of $\mathfrak{A}$ must be activated in state s_0 precisely at the nodes with $v(x) = \tau$, this copy will always have the highest number. Recall that we assume $\mathfrak{A}$ to be such that s_0 can appear only once in an $\mathfrak{A}$-run. If d is avoided, then each active copy i, $i < k$, of $\mathfrak{A}$ independently moves into the state $w_1(i)$ at $x0$ and $w_2(i)$ at $x1$ where $(w_1(i), w_2(i)) \in M(u(i), \sigma)$.

Now let $w_1(i_1), \ldots, w_1(i_{p-1})$, $i_1 < i_2 < \ldots < i_{p-1}$ be all the pairwise different states in $w_1(0), \ldots, w_1(k-1)$, where we define $i_j = \min(i \mid w_1(i) = w_1(i_j))$. If (u_1, ϕ_1) is the state of $\mathfrak{B}$ at $x0$, then $l(u_1) = p$ or $l(u_1) = p + 1$ (in which case $u_1(p) = s_0$), and $u_1(j) = w_1(i_j)$, $j < p$. Also, $\phi_1\colon k \to p$ and $\phi_1(i) = j$ if and only if $u_1(i) = w_1(i_j)$. Similar statements hold for the state $(u_2, \phi_2) \in S$ at $x1$.

We introduce a set $S_0^{\mathfrak{B}} = \{(\Lambda, \emptyset), (s_0, \emptyset)\}$ of initial states of $\mathfrak{B}$.

DEFINITION 10. Let $t = (v, T) \in V_\Sigma$, and $r \in \mathrm{Rn}(\mathfrak{B}, t)$ be such that $r(\Lambda) \in S_0^{\mathfrak{B}}$ and $r(x) \neq d$, $x \in T$. Denote $r(x) = (u^x, \phi^x)$, $x \in T$. If $l(u_x) = k$, $m < k$, then we shall say that m is *active at* x; if $u^x(k-1) = s_0$ then $k-1$ is *activated at* x. (Note that $k-1$ is activated at x if and only if $v(x) = \tau$.)

Let m be active at x and $y = x0$ or $y = x1$. We say that *m at x is replaced by m_1 at y* $((m,x) \to (m_1, y))$ if $m_1 = \phi^y(m)$. The notion of replacement is extended by passing to the transitive closure. Thus assume $x < y$, $x = x_0$, $x_{i+1} = x_i\varepsilon_i$, $\varepsilon_i \in \{0,1\}$, $0 \leqq i \leqq k-1$, and $y = x_k$. We shall say that *m at x is replaced by m' at y* $((mx) \to (m', y))$, if for a sequence $m_i \in n$, $0 \leqq i \leqq k$, $m_0 = m$, $m_k = m'$, and $(m_i, x_i) \to (m_{i+1}, x_{i+1})$, $0 \leqq i \leqq k-1$.

Henceforth, to the end of this section, t, r, and (u^x, ϕ^x), retain their above meanings.

LEMMA 10. *Let k be activated at x. Define $r_x : T_x \to S$ by $r_x(x) = s_0 = u^x(k)$, $r_x(y) = u^y(m)$ if $x < y$ and $(k,x) \to (m,y)$. For all x such that $v(x) = \tau$, $r_x \in \mathrm{Rn}(\mathfrak{A}, (v, T_x))$. The proof is obvious.*

Let $\pi \subset T$ be a path, and let $x \in \pi$. We say that m is *stabilized at x* (along π) if m is active at x and $(m,x) \to (m,y)$ for all $x < y \in \pi$.

LEMMA 11. *Let k be active at $z \in \pi$, there exists a $z \leqq x \in \pi$ and an m such that $(k,z) \to (m,x)$ and m is stabilized at x along π. If m is stabilized at $x \in \pi$ along π and $m_1 \leqq m$ then m_1 is stabilized at x along π. For every path $\pi \subset T$ for which $r(\pi) \neq \{(\Lambda, \emptyset)\}$ (i.e. $\tau \in v(\pi)$) there exists a $x_0 \in \pi$ and an m_0 such that m_0 is stabilized at x_0 and every $m > m_0$ is not stabilized at any $x \in \pi$.*

Proof. The statements follow at once from the properties of the contracting function mentioned just before Definition 9.

LEMMA 12. *$t \in D$ if and only is there exists a run $r \in \mathrm{Rn}(\mathfrak{B}, t)$ such that $r(\Lambda) \in S_0^{\mathfrak{B}}$, $d \notin r(T)$, and for every $\pi \subset T$ if m is stabilized at $z \in \pi$ along π, then $c(\{y \mid z \leqq y \in \pi,\ u^y(m) \in F\}) = \omega$.*

Proof. Assume the existence of such a run r. Since $d \notin r(T)$, a $k < n$ is activated at precisely those x for which $v(x) = \tau$. Now $r_x \in \mathrm{Rn}(\mathfrak{A}, (v, T_x))$, by Lemma 10, and $r_x(x) = s_0$. Let $\pi \subset T_x$ be a path. There exists a $x < z \in \pi$, a $m \leqq k$ such that $(k,x) \to (m,z)$ and m is stabilized at z along π. For $z \leqq y \in \pi$ we have $r_x(y) = u^y(m)$. Hence $\mathrm{In}(r_x \mid \pi) \cap F \neq \emptyset$. This implies $(v, T_x) \in T(\mathfrak{A})$. Thus $t \in D$.

Assume now $t \in D$. For every x such that $v(x) = \tau$ we have $(v, T_x) \in T(\mathfrak{A})$. Let $\bar{r}_x \in \mathrm{Rn}\, \mathfrak{A}, (v, T_x))$ be an accepting run.

Let $r: T \to S^{\mathfrak{B}}$ be a $\mathfrak{B}$-run on t. For m active at $x \in T$, define $\rho(m,x)$- the *origin* of m at x, by: $\rho(m,x) = x$ if m is activated at x; and $\rho(m,x) = y$ if $y < x$ and for some k, $(k,y) \to (m,x)$, and no $z < y$ has this property. It is readily verified that $v(\rho(m,x)) = \tau$ holds for every $x \in T$ and m active at x. Note that $\rho(m,x)$ is completely determined by $r \big| \{y \big| y \leqq x\}$.

We shall define by induction on $l(x)$, $x \in T$, a run $r \in \mathrm{Rn}(\mathfrak{B}, t)$. Put $r(\Lambda) = (s_0, \emptyset)$ if $v(\Lambda) = \tau$; $r(\Lambda) = (\Lambda, \emptyset)$ otherwise. Assume that $r(x) = (u^x, \phi^x)$ has been defined for all x such that $l(x) \leqq h$ and that $u^x(m) = \bar{r}_{\rho(m,x)}(x)$, $m < l(u^x)$, for all these x. Let $l(x) = h$, $l(u^x) = k$. Put $w_{a+1}(m) = \bar{r}_{\rho(m,x)}(xa)$, $m < k$, $a = 0,1$. Since $\bar{r}_{\rho(m,x)}$ is an $\mathfrak{A}$-run on $(v, T_{\rho(m,x)})$, and $u^x(m) = \bar{r}_{\rho(m,x)}(x)$, it follows that $(w_1(m), w_2(m)) \in M(u^x(m), v(x))$.

Let $y = x0$ (the case $y = x1$ is treated in the same way). Define $u^y = C(w_1(0) \dots w_1(k-1))$ if $v(y) \neq \tau$, and let ϕ^y be the corresponding contracting function. Define $u^y = C(w_1(0) \dots w_1(k-1))s_0$ if $v(y) = \tau$. Obviously $((u^{x0}, \phi^{x0}), (u^{x1}, \phi^{x1})) \in M^{\mathfrak{B}}((u^x, \phi^x), v(x))$. If $m = \min(m' \big| \phi^y(m') = i)$ then $\rho(i,y) = \rho(m,x)$. Thus $u^y(i) = w_1(m) = \bar{r}_{\rho(m,x)}(y) = \bar{r}_{\rho(i,y)}(y)$. Thus r was extended to a run $r: \{x \big| l(x) \leqq h + 1\} \to S^{\mathfrak{B}}$ with the same properties.

In this way we obtain a run $r \in \mathrm{Rn}(\mathfrak{B}, t)$ such that $d \notin r(T)$ and $u^y(m) = \bar{r}_{\rho(m,y)}(y)$ for $y \in T$, and m active at y. If $\pi \subset T$ and m is stabilized at $z \in \pi$, then $\rho(m,z) = x \in \pi$ and, for $z \leqq y$, $\rho(m,z) = \rho(m,y) = x$. Hence $u^y(m) = \bar{r}_x(y)$ for $z \leqq y$. Since $\mathrm{In}(\bar{r}_x \big| \pi) \cap F \neq \emptyset$, also $c(\{y \big| u^y(m) \in F,\ z \leqq y \in \pi\}) = \omega$.

The proof of Theorem 9 will result from the following.

LEMMA 13. *Let R be the set of $S^{\mathfrak{B}}$-valued trees (r, T) such that* a) $\exists t[r \in \mathrm{Rn}(\mathfrak{B}, t) \wedge t \in V_\Sigma]$; b) $d \notin r(T)$, $r(\Lambda) \in S_0^{\mathfrak{B}}$; c) *for every path $\pi \subset T$, $x \in \pi$ and m which is stabilized at x, $c(\{y \big| x < y \in \pi, u^y(m) \in F\}) = \omega$. The set R is s.f.a. definable.*

Proof. The sets R_a and R_b of $S^{\mathfrak{B}}$-trees satisfying conditions a) and b) are certainly definable by s.f.a. Thus it will be enough to construct an automaton $\mathfrak{C}$ accepting a $(r, T) \in R_a \cap R_b$ if and only if it satisfies c). The set of states of $\mathfrak{C}$ will be $n + 1 = \{0, 1, \dots, n\}$. Define $\bar{M}(n, s) = (0, 0)$, $s \in S^{\mathfrak{B}}$; $\bar{M}(m, (u, \phi)) = (m+1, m+1)$, for $m < n$, $(u, \phi) \in S^{\mathfrak{B}}$, if $l(u) \leqq m$ or $\phi(m) < m$ or $u(m) \in F$; $\bar{M}(m, u, \phi)) = m$ otherwise. Put $\mathfrak{C} = \langle n+1, \bar{M}, 1, \{n\} \rangle$. If $h: T \to n + 1$, $h(\Lambda) = 0$ is the $\mathfrak{C}$-run on a tree $(r, T) \in R_a \cap R_b$ and $\pi \subset T$ then, by use of Lemma 11, we see that $n \notin \mathrm{In}(h \big| \pi)$ if and only if for some $x \in \pi$ and m stabilized at x, $c(\{y \big| x < y \in \pi,\ u^y(m) \in F\}) < \omega$. Thus $R = R_a \cap R_b \cap T(\mathfrak{C})$.

Proof of Theorem 9. By Lemma 12, $t = (v, T) \in D$ if and only if there exists a run $r \in \mathrm{Rn}(\mathfrak{B}, t)$ such that $(r, T) \in R$. Consider the alphabet

$\bar{\Sigma} = S^{\mathfrak{B}} \times \Sigma$. The set A of $\bar{\Sigma}$-trees $(\bar{v}, T)$ such that $p_0\bar{v} \in \mathrm{Rn}(\mathfrak{B},(p_1\bar{v}, T))$, is s.f.a. definable. The set $B = \{(\bar{v}, T) | (p_0\bar{v}\, T) \in R\}$ is s.f.a. definable by Lemma 13 and Theorem 3. Now $D = p_1(A \cap B)$.

5. Automata on finite trees

We shall need some notions concerning finite Σ-trees and the definition of sets of such trees by automata. A *finite* Σ-tree is a pair (v, E) where $E \subset T$ is a finite frontiered tree and $v: E - Ft(E) \to \Sigma$. An automaton on finite Σ-trees is a system $\mathfrak{A} = \langle S, M, s_0, f \rangle$ where $M: (S - \{f\}) \times \Sigma \to P(S \times S)$, and $s_0 \in S$. The notion of an $\mathfrak{A}$-run on a finite Σ-tree is defined in the obvious way. An automaton $\mathfrak{A} = \langle S, M, s_0, f \rangle$ *accepts* $e = (v, E)$ if there exist an accepting $\mathfrak{A}$-run r on e such that $r(\Lambda) = s_0$ and $r(Ft(E)) = \{f\}$. The set of all the e accepted by $\mathfrak{A}$ is denoted by $T_f(\mathfrak{A})$. In [1, 5, 7] the notion of automata on finite trees is defined in a different way. It is, however, not hard to show the equivalence of the various definitions.

THEOREM 14. *Let $\mathfrak{A} = \langle S, M, s_0, f \rangle$ be a Σ-automaton. Define a set $B \subseteq V_\Sigma$ by: $(v, T) \in B$ if and only if there exists a sequence $(E_n)_{n<\omega}$ of frontiers of T such that $E_n < E_{n+1}$ and $(v, G_n) \in T_f(\mathfrak{A})$, $n < \omega$, where $G_n = \{x | x \leqq y$ for some $y \in E_n\}$. The set B is s.f.a. definable.*

Proof. Define a s.f.a. $\mathfrak{B} = \langle \bar{S}, \bar{M}, \bar{s}_0, \bar{F} \rangle$ as follows. Put $\bar{S} = S \times S$; $\bar{s}_0 = (s_0, f)$; $\bar{F} = S \times \{f\}$. The table $\bar{M}$ is defined by $((s_1, s_1'), (s_2, s_2')) \in \bar{M}((s, s'), \sigma)$ if and only if $(s_1, s_2) \in M(s, \sigma)$, $(s_1', s_2') \in M(s', \sigma)$ for $s' \neq f$, and $(s_1', s_2') \in M(s, \sigma)$ for $s' = f$. We claim $T(\mathfrak{B}) = B$.

Let $t = (v, T) \in T(\mathfrak{B})$ and let $\bar{r}: T \to \bar{S}$ be an accepting $\mathfrak{B}$-run. There exists a sequence $(E_n)_{n<\omega}$ of frontiers of T such that $\bar{r}(E_n) \subseteq \bar{F}$, $E_n < E_{n+1}$, $n < \omega$. Let $G_n = \{x | x \leqq y$ *for some* $y \in E_n\}$, $n < \omega$. Consider the mapping $r': G_{n+1} \to S$ defined by: $r'(x) = p_0(\bar{r}(x))$ for $x \in G_n$, and $r'(x) = p_1(\bar{r}(x))$ for $x \in G_{n+1} - G_n$. Clearly r' is an $\mathfrak{A}$-run on (v, G_{n+1}) and $r'(E_{n+1}) = \{f\}$. Thus $(v, G_{n+1}) \in T_f(\mathfrak{A})$, $n < \omega$. Hence $(v, T) \in B$.

Assume $(v, T) \in B$. Let $(E_n)_{n<\omega}$ and $(v, G_n)_{n<\omega}$ be as in the statement of the theorem. Let $r_n: G_n \to S$, $n < \omega$, be an $\mathfrak{A}$-run on (v, G_n) so that $r_n(E_n) = \{f\}$. Define $T_n = \{x | l(x) = n\}$. By König's Lemma, there exists a sequence $(n(i))_{i<\omega}$ of integers such that $n(i) < n(i+1)$, $i < \omega$, and $r_{n(i)} | T_i = r_{n(j)} | T_i$ for $i \leqq j < \omega$. Denote $r_i' = r_{n(i)}, E_i' = E_{n(i)}, G_i' = G_{n(i)}$, $i < \omega$. Let $r: T \to S$ be the limiting mapping of the sequence $(r_i')_{i<\omega}$, i.e. $r(x) = r_i'(x)$ for x such that $l(x) \leqq i$. r is an $\mathfrak{A}$-run satisfying $r(\Lambda) = s_0$.

We shall define inductively a sequence $m(k)$, $k < \omega$, of integers and $\mathfrak{B}$-runs $\bar{r}_k$: $G_{m(k)}' \to \bar{S}$. Define $m(0) = 0$, $p_0\bar{r}_0 = r | G_0'$, $p_1\bar{r}_0(x) = r_0'(x)$

for $\Lambda < x \in G_0'$, and $\bar{r}_0(\Lambda) = (s_0, f) = \bar{s}_0$. Assume $(m(i))_{i \leqq k}$ and $(\bar{r}_i)_{i \leqq k}$ to be defined. Let $j = \max(l(x) \mid x \in E'_{m(k)})$. Then $E'_{m(k)} < E'_j$ and $r'_j \mid G'_{m(k)} = r \mid G'_{m(k)}$. Define $m(k+1) = j$; $p_0 \bar{r}_{k+1} = r \mid G'_j$; $p_1 \bar{r}_{k+1}(x) = r'_j(x)$ for $x \in G'_j - G'_{m(k)}$. The sequence $(\bar{r}_k)_{k<\omega}$ defined in this way has the properties that $\bar{r}_{k+1} \mid G'_{m(k)} = \bar{r}_k$, and $p_1 \bar{r}_k(E_{m(k)}) = \{f\}$, $k < \omega$. The run $\bar{r} = \lim_{k \to \infty} \bar{r}_k$ is an accepting $\mathfrak{B}$-run of (v, T), hence $(v, T) \in T(\mathfrak{B})$. Thus $B = T(\mathfrak{B})$.

6. **Weak definability and s.f.a.**

In this section let $\Sigma = \{0, 1\}$. χ_A will denote the characteristic function of A. Let $A = (A_1, \ldots, A_n)$ be an n-tuple of subsets of T. With A we associate the Σ^n-tree $\tau(A) = (v_A, T)$ where $v_A(x) = (\chi_{A_1}(x), \ldots, \chi_{An}(x))$, $x \in T$. The mapping $\tau: P(T)^n \to V_{\Sigma^n, \Lambda}$ is clearly a one-to-one correspondence.

Definition 11. Let $R \subseteq P(T)^n$. We say that the Σ^n-automaton $\mathfrak{A}$ *represents* R if

$$(2) \qquad \tau(R) = T(\mathfrak{A}) \cap V_{\Sigma_n, \Lambda}.$$

Lemma 15. *Let $K_1 \subseteq P(T)^n$ be the set of all $A = (A_1, \ldots, A_n)$ such that $c(A_1) = 1$; and let $K_2 \subseteq P(T)^n$ be the set of all A such that $c(A_1) < \omega$. The sets K_1 and K_2 are representable by special automata.*

Proof. Since $\Sigma = \Sigma \times \Sigma^{n-1}$ it suffices, by Theorem 3, to consider the case $n = 1$.

Let $\mathfrak{A}_1 = \langle \{s_0, s_1, s_2\}, M_1, s_0, \{s_1\} \rangle$, where $M_1(s_0, 0) = \{(s_0, s_1), (s_1, s_0)\}$, $M_1(s_0,1) = M_1(s_1,0) = \{(s_1,s_1)\}$, and $M_1(s_1,1) = M_1(s_2,0) = M_1(s_1,1) = \{(s_2,s_2)\}$.

Let $\mathfrak{A}_2 = \langle \{s_0, s_1, s_2\}, M_2, s_0, \{s_1\} \rangle$, where $M_2(s_0, 1) = M_2(s_0, 1) = \{(s_0,s_0),(s_1,s_1)\}$, $M_2(s_1,0) = \{(s_1,s_1)\}$, $M_2(s_1,1) = M_2(s_2,0) = M_2(s_2,1) = \{(s_2,s_2)\}$.

It can easily be verified that K_i is represented by $\mathfrak{A}_i$, $i = 1, 2$.

Definition 12. Let $A = (x_1, \ldots, x_m, \alpha_{m+1}, \ldots, \alpha_{m+n}, A_{m+n+1}, \ldots, A_{m+n+k})$, where $x_i \in T$ $1 \leqq i \leqq m$, $\alpha_i \subseteq T$ is finite, $m+1 \leqq i \leqq m+n$, and $A_i \subseteq T$ $m+n+1 \leqq i \leqq m+n+k$. We shall say that A is of *type* (m, n, k). Note that some of the A_i may also be finite but that we disregard this in defining the type. We shall represent A by $\sigma(A) = (A_i)_{1 \leqq i \leqq m+n+k}$ where $A_i = \{x_i\}$, $1 \leqq i \leqq m$, $A_i = \alpha_i$, $m+1 \leqq i \leqq m+n$, and the last k terms of $\sigma(A)$ coincides with those of A.

Corollary 16. *Let $K(m,n,k) = \{\sigma(A) \mid A$ is of type $(m,n,k)\}$. The set $K(m,n,k)$ is representable by a special automaton.*

Proof. This follows from Lemma 15 and from the closure of s.f.a. definable sets under intersections (Theorem 2).

THEOREM 17. *Let $R \subseteq P(T)^n$ and $Q \subseteq P(T)^n$ be, respectively, represented by the special automata $\mathfrak{A}$ and $\mathfrak{B}$. The following sets are representable by special automata.*

(a) $R \cup Q$;

(b) $R \cap Q$;

(c) $R_1 = \{(A_1, \dots, A_{n-1}) \mid \exists A_n[(A_1, \dots, A_{n-1}) \in R]\}$;

(d) $R_2 = \{(A_1, \dots, A_{n-1}) \mid \forall \alpha_n[(A_1, \dots, A_{n-1}, \alpha_n) \in R]\}$,

here α_n ranges over all finite subsets of T.

Proof. Since $\tau(R \cup Q) = \tau(R) \cup \tau(Q)$, $\tau(R \cap Q) = \tau(R) \cap \tau(Q)$, (a) and (b) follow from Theorem 2.

Let p be the mapping $(x_1, \dots, x_{n-1}, x_n) \to (x_1, \dots, x_{n-1})$. We have $R_1 = pR$ Now $\Sigma^n = \Sigma^{n-1} \times \Sigma$ and p, which may also be considered as inducing the projection $p_0 : \Sigma^{n-1} \times \Sigma \to \Sigma^{n-1}$, commutes with τ. I.e. for $A \in P(T)^n$, $\tau(p(A)) = p(\tau(A))$. Thus

$$\tau(R_1) = \tau(p(R)) = p(\tau(R)) = p(T(\mathfrak{A}) \cap V_{\Sigma_n, \Lambda});$$

and R_1 is representable by Theorem 3.

Let (v, T) be a Σ^{n-1}-tree and $\chi : T \to \{0, 1\}$. By $(v \times \chi, T)$ we shall denote the $\Sigma^n = \Sigma^{n-1} \times \Sigma$-tree such that $(v \times \chi)(x) = v(x)\chi(x)$, $x \in T$. A similar notation will be used for finite trees (v, E).

To prove (d), let us look at $\bar{\Sigma} = \Sigma^{n-1} \times (P(S) - \{\emptyset\})$-trees, where S is the set of states of $\mathfrak{A} = \langle S, M, s_0, F \rangle$. Let η be the mapping $\eta : T \to \{0\}$. Define $\bar{A} \subseteq V_{\bar{\Sigma}}$ to be the set of trees $(\bar{v}, T_x)$ so that for each $s \in p_1 \bar{v}(x)$, $((p_0 \bar{v}) \times \eta, T_x)$ is accepted by $\langle S, M, s, F \rangle$. The set A is obviously s.f.a. definable.

Define now an invariant set $A \subseteq V_{\bar{\Sigma}}$ by $(\bar{v}, T) \in A$ if and only if $(\bar{v}, T_x) \in \bar{A}$ for *all* $x \in T$. It follows at once from Theorem 9 that A is s.f.a. definable.

Consider the following set P of finite $\bar{\Sigma}$-trees. $(\bar{v}, G) \in P$ if and only if there exists a frontiered tree $H \subset G$ such that 1) for every $y \in Ft(G)$ there exists an $x \in Ft(H)$ such that $y = x0$ or $y = x1$; 2) for every $\chi : H \to \{0, 1\}$ such that $\chi(Ft(H)) = \{0\}$, there exists an $\mathfrak{A}$-run $r : H \to S$ on $((p_0 \bar{v}) \times \chi, H)$ such that $r(\Lambda) = s_0$ and $r(x) \in p_1 \bar{v}(x)$ for all $x \in Ft(H)$. The above conditions for $(\bar{v}, G) \in P$ can be expressed in the weak second-order theory of two successor functions. Hence $P = T_f(\mathfrak{C})$ for some finite automaton $\mathfrak{C}$ (see [1, 7]).

Thus, by Theorem 14, the invariant set $B \subset V_{\bar{\Sigma}}$ defined by $(\bar{v}, T) \in B$ if and only if there exists a sequence $(G_n)_{n<\omega}$ of finite subtrees of T such that $Ft(G_n) < Ft(G_{n+1})$ and $(\bar{v}, G_n) \in P$, $n < \omega$, is definable by a special automaton.

If we shall show that $p_0(A \cap B) \cap V_{\Sigma^{n-1}, \Lambda} = \tau(R_2)$, then the proof of (d) will be completed by Theorem 3.

Let $t = (v, T)$ be a Σ^{n-1}-tree such that $t = \tau((A_1, \ldots, A_{n-1}))$. Assume $t \in p_0(A \cap B)$. Then there exists a $\bar{\Sigma} = \Sigma^{n-1} \times (P(S) - \emptyset)$-tree $\bar{t} = (\bar{v}, T)$ such that $p_0\bar{v} = v$, and $\bar{t} \in A \cap B$. Let $\alpha \subset T$ be a finite set. $\tau((A_1, \ldots, A_{n-1}, \alpha)) = (v \times \chi_\alpha, T)$. Now $(\bar{v}, T) \in B$, hence there exist finite trees $H \subset G$ such that conditions 1–2 of the definition of P are satisfied, and for $x \in (T - H) \cup Ft(H)$, $\chi_\alpha(x) = 0$. Namely, with the notations of the previous paragraph, $G = G_n$ for some large enough $n < \omega$. Thus there exists a $r \in \mathrm{Rn}(\mathfrak{A}, (v \times \chi_\alpha, H))$ such that $r(\Lambda) = s_0$ and $r(x) \in p_1\bar{v}(x)$ for $x \in Ft(H)$. Since $(\bar{v}, T) \in A$, and $\chi_\alpha(y) = 0$ for $y \in T_x$ and $x \in Ft(H)$, $r(x) \in p_1\bar{v}(x)$ implies that $(v \times \chi_\alpha, T_x)$ is accepted by $\langle S, M, r(x), F \rangle$ for $x \in Ft(H)$; let $r_x : T_x \to S$ be an accepting run. Define $r' : T \to S$ by $r'(y) = r(y)$ for $y \in H$, $r'(y) = r_x(y)$ for $y \in T_x$, $x \in Ft(H)$. Since $r_x(x) = r(x)$ for $x \in Ft(H)$, the mapping r' is well-defined. It is readily seen that r' is an accepting run of $\mathfrak{A}$ on $(v \times \chi_A, T)$. Hence $\tau((A_1, \ldots, A_{n-1}, \alpha)) \in T(\mathfrak{A})$ for every finite α. Thus $(A_1, \ldots, A_{n-1}) \in R_2$.

Assume now that $t = (v, T) = \tau((A_1, \ldots, A_{n-1}))$, where $(A_1, \ldots, A_{n-1}) \in R_2$. Let $\eta : T \to \{0, 1\}$, $\eta(T) = \{0\}$. For every $x \in T$ define $s(x) = \{s \mid s \in S, (v \times \eta, T_x) \in T(\langle S, M, s, F \rangle)$. Since $\eta = \chi_\emptyset$ and $(A_1, \ldots, A_{n-1}, \emptyset) \in R$, it follows that $s(x) \neq \emptyset$ for $x \in T$. Let $\bar{t} = (\bar{v}, T)$ be the $\bar{\Sigma}$-tree such that $p_0\bar{v} = v$ and $p_1\bar{v}(x) = s(x)$ for all $x \in T$. We want to show that $\bar{t} \in A \cap B$ which will entail $t \in p_0(A \cap B)$ and hence $\tau(R_2) = p_0(A \cap B) \cap V_{\Sigma^{n-1}, \Lambda}$.

That $\bar{t} \in A$ follows at once from the definition of $s(x)$ and $\bar{v}$. Now let $H \subset T$ be any finite tree and $G = H \cup \{xa \mid x \in Ft(H),\ a = 0, 1\}$. Let $\chi : H \to \{0, 1\}$ be such that $\chi(Ft(H)) = \{0\}$, and let $\alpha \subset T$ be the finite set such that $\chi_\alpha \mid H = \chi$, $\chi_\alpha(y) = 0$ for $y \notin H$. Since $(A_1, \ldots, A_{n-1}, \alpha) \in R$, it follows that $(v \times \chi_\alpha, T) \in T(\mathfrak{A})$. Let $r \in \mathrm{Rn}(\mathfrak{A}, (v \times \chi_\alpha, T))$ be an accepting run. Then $r \in \mathrm{Rn}(\mathfrak{A}, (v \times \chi, H))$, $r(\Lambda) = s_0$. Now $(v \times \chi_\alpha, T_x)$ is accepted by $\langle S, M, r(x), F \rangle$, $x \in Ft(H)$. Since $\chi_\alpha(y) = 0$ for $y \in T_x$, $x \in Ft(H)$, we have $r(x) \in s(x)$. Thus $(\bar{v}, G) \in P$. Chose now $G_n = \{x \mid l(x) \leqq n + 1\}$, $n < \omega$, then $(\bar{v}, G_n) \in P$. Hence $\bar{t} \in B$.

Lemma 18. *Let $w \in \{0, 1\}^*$ and $P \subseteq P(T)^n$ be the set of n-tuples $(A_1, \ldots, A_n)$ such that $A_i = \{x\}$ for some $x \in T$ and $xw \in A_j$. Let Q be similarly defined with $xw \notin A_j$. The relations P and Q are representable by special automata.*

Proof. Because of Theorem 3 it suffices to consider the case $n = 2$, $i = 1$, $j = 2$. We shall construct a Σ^2-automaton. Let $w = \varepsilon_1 \ldots \varepsilon_k$. Put $S = \{s_0, s_1, \ldots, s_k, d\}$, (if $w = \Lambda$ then $S = \{s_0, d\}$).

Let $\delta = 0, 1$, $\sigma \in \Sigma^2$. Define $M(s_0, (0, \delta)) = \{(s_0, s_0)\}$, $M(s_0, (1, \delta)) = \{(s_1, s_0)\}$ if $\varepsilon_1 = 0$, and $M(s_0, (1\ \delta)) = \{(s_0, s_1)\}$ if $\varepsilon_1 = 1$. For $1 \leqq i < k$, define $M(s_i, \sigma) = \{(s_{i+1}, s_0)\}$ if $\varepsilon_\iota = 0$, $M(s_i, \sigma) = \{(s_0, s_{i+1})\}$ if $\varepsilon_{i+1} = 1$.

Finally, $M(s_k,(\delta,1)) = \{(s_0,s_0)\}$, $M(s_k,(\delta,0)) = M(d,\sigma) = \{(d,d)\}$. It is easily verified that $V_{\Sigma^2,\Lambda} \cap T(\langle S, M, s_0, \{s_0\}\rangle) \cap K(1,0,1) = \tau(P)$. It follows from Corollary 16 that P is representable by a s.f.a.

The proof for Q is as above with an appropriate change in the definition of $M(s_k,\sigma)$.

THEOREM 19. *If* $F(\boldsymbol{x},\boldsymbol{\alpha},\boldsymbol{A})$, *where* $\boldsymbol{x} = (\boldsymbol{x}_1,\ldots,\boldsymbol{x}_n)$, $\boldsymbol{\alpha} = (\boldsymbol{\alpha}_{m+1},\ldots,\boldsymbol{\alpha}_{m+n})$, $\boldsymbol{A} = (\boldsymbol{A}_{m+n+1},\ldots,\boldsymbol{A}_{m+n+k})$, is a formula of S2S involving quantifiers over individual variables and finite-set variables only, and

$$P = \{(x,\alpha,A) \mid \mathcal{N}_2 \models F(x,\alpha,A)\},$$

then $\sigma(P) = R(F)$ (see Definition 12) is representable by a s.f.a.

Proof. In F replace all occurrences of $t_1 = t_2$, where t_1, t_2 are terms, by $\forall\boldsymbol{\alpha}[t_1 \in \boldsymbol{\alpha} \rightarrow t_2 \in \boldsymbol{\alpha}]$. Replace all occurrences of $u = v$, where u and v are finite or arbitrary set variables, by $\forall\boldsymbol{x}[\boldsymbol{x} \in u \leftrightarrow \boldsymbol{x} \in v]$. Call a Boolean (propositional) combination of formulas *positive* if it involves only the use of $\vee$ and $\wedge$. Transform F into an equivalent formula by transporting the negation sign inside next to the atomic formulas. [1]

This is done by use of the rules $\sim[F \vee G] \equiv [\sim F \wedge \sim G]$, $\sim[F \wedge G] \equiv [\sim F \vee \sim G]$, $\sim\forall vF \equiv \exists v \sim F$, $\sim\exists vF \equiv \forall v \sim F$, where v is an individual or finite-set variable. Thus we may assume that F is obtained from atomic formulas $t \in u$ and $\sim t \in u$, where t is a term and u a set variable, by forming positive boolean combinations and using quantifiers over individual and finite-set variables.

The proof is by induction on the structure of F. If F is in one of the forms $x_i w \in A_j$, $x_i w \in \alpha_j$, where $w \in T$, or a negation of such a formula, then $R(F)$ is representable by Lemma 18, Corollary 16, and Theorem 3.

If $R(F_1)$ and $R(F_2)$ are representable then, since $R(F_1 \vee F_2) = R(F_1) \cup R(F_2)$ and $R(F_1 \wedge F_2) = R(F_1) \cap R(F_2)$, so are $R(F_1 \vee F_2)$ and $R(F_1 \wedge F_2)$ by Theorem 17(a), (b).

Let $F(\boldsymbol{x},\boldsymbol{\alpha},\boldsymbol{A})$ be a formula of the form in the statement of the theorem, then $R(F)$ is of type (m,n,k). Assume that $F(F)$ is representable and let v_i be one of the individual or finite-set variables of F; i.e. $i \leq m+n$. We have

$$(3)\quad R(\exists v_i F) = \{(A_1,\ldots,A_{i-1},A_{i+1},\ldots,A_{m+n+k}) \mid \exists A_i[(A_1,\ldots,A_i,\ldots,A_{n+m+k}) \in R(F)]\}$$

$$(4)\quad R(\forall v_i F) = \{(A_1,\ldots,A_{i-1},A_{i+1},\ldots,A_{m+n+k}) \mid \forall\alpha_i[(A_1,\ldots,A_{i-1},\alpha_i,A_{i+1},\ldots,A_{m+n+k}) \in R(F)\},$$

[1] The idea of pushing negation signs next to the atomic formulas, was suggested to me by M. Magidor.

here $\alpha \subset T$ ranges over finite sets. These relations (3) and (4) hold because $R(F) \subseteq K(m,n,k)$, so that in (3) if $i \leqq m$, A_i will automatically range over singletons and if $m < i \leqq m+n$ then A_i will range over finite sets; in (4) for $i \leqq m+n$, any A_i for which $(A_1, \cdots, A_i, \cdots, A_{m+n+k}) \in R(F)$ is finite so that quantifying over just finite sets is no restriction. Now $R(\exists v_i F)$ and $R(\forall v F)$ are representable by Theorem 17(c), (d).

Thus our proof is completed by induction on formulas.

LEMMA 20. *The partial order $x \leqq y$ on T, and the lexicographic order $x \preccurlyeq y$ defined by $x \preccurlyeq y \equiv x \leqq y \vee \exists z[z0 \leqq x \wedge z1 \leqq y]$, are weakly definable.*

Proof. The formula

$$\forall \boldsymbol{\alpha}[\forall \boldsymbol{u} \forall \boldsymbol{v}[\boldsymbol{u} \in \boldsymbol{\alpha} \wedge [\boldsymbol{v}0 = \boldsymbol{u} \vee \boldsymbol{v}1 = \boldsymbol{u}] \rightarrow \boldsymbol{v} \in \boldsymbol{\alpha}] \wedge \boldsymbol{y} \in \boldsymbol{\alpha} \rightarrow \boldsymbol{x} \in \boldsymbol{\alpha}],$$

defines $x \leqq y$. That $x \preccurlyeq y$ is definable, follows from the weak definability of $x \leqq y$.

LEMMA 21. *The relations $\alpha \subset T_x$ is a frontier of T_x; $\beta \subset T$ is a finite frontiered tree; and $\alpha = Ft(\beta)$; are weakly definable.*

Proof. The formula

$$\forall \boldsymbol{y}[\boldsymbol{y} \in \boldsymbol{\alpha} \rightarrow \boldsymbol{x} \leqq \boldsymbol{y}] \wedge \forall \boldsymbol{y} \forall \boldsymbol{z}[\boldsymbol{y} \in \boldsymbol{\alpha} \wedge \boldsymbol{z} \in \boldsymbol{\alpha} \wedge \boldsymbol{y} \leqq \boldsymbol{z} \rightarrow \boldsymbol{y} = \boldsymbol{z}] \wedge$$
$$\forall \boldsymbol{y} \exists \boldsymbol{z}[\boldsymbol{x} \leqq \boldsymbol{y} \rightarrow \boldsymbol{z} \in \boldsymbol{\alpha} \wedge [\boldsymbol{y} \leqq \boldsymbol{z} \wedge \boldsymbol{z} \leqq \boldsymbol{y}]],$$

defines the relation: $\alpha \subset T_x$ and α a frontier of T_x. The other statements are immediate.

THEOREM 22. *A relation $R \subseteq P(T)^n$ is representable by a s.f.a. if and only if for some m there exists a weakly definable relation $P \subseteq P(T)^{n+m}$ such that*

$$R = \{(A_1, \ldots, A_n) \,|\, \exists B_1 \ldots \exists B_m[(A_1, \ldots, A_n, B_1, \ldots, B_m) \in P]\}. \tag{5}$$

Proof. That a relation R satisfying (5) for a weakly definable P is representable, follows from Theorem 19 and Theorem 17(c).

Assume $\tau(R) = T(\mathfrak{A}) \cap V_{\Sigma^n, \Lambda}$, where $\mathfrak{A} = \langle S, M, s_0, F \rangle$. By possibly adding states to S, we may assume $S = \{0,1\}^m$ for some $m < \omega$. Thus every $r = (B_1, \ldots, B_m) \in P(T)^m$ may be viewed as the run $r: T \rightarrow S$ such that $r(x) = (\chi_{B_1}(x), \ldots, \chi_{Bm}(x))$, $x \in T$. This sets a 1-1 correspondence between $P(T)^m$ and the set S^T.

It is readily checked that the relations $r(x) = s$, where $s \in S$, $x \in T$, and $r(x) \in F \subseteq S$, are weakly definable. Hence $r \in \mathrm{Rn}(\mathfrak{A}, \tau(A_1, \ldots, A_n)) \wedge r(\Lambda) = s_0$, where $r = (B_1, \ldots, B_m)$, is weakly definable (by a formula H involving quantification over just individual variables). The relation $\forall\pi[\mathrm{In}(r|\pi)\cap F \neq \phi]$ is shown to be weakly definable by transforming

$$\forall x \exists \alpha[\alpha \textit{ frontier of } T_x \textit{ and } \forall y[y \in \alpha \to r(y) \in F]]$$

into a formula $G(\boldsymbol{B}_1, \ldots, \boldsymbol{B}_m)$ of S2S which involves quantification over just individual variables and the quantifier $\exists\boldsymbol{\alpha}$. Now $\tau(A_1, \ldots, A_n) \in T(\mathfrak{A})$, which is equivalent to $(A_1, \ldots, A_n) \in \mathrm{R}$, defined in S2S by

$$\exists \boldsymbol{B}_1 \ldots \exists \boldsymbol{B}_m[H(\boldsymbol{A}_1, \ldots, \boldsymbol{A}_n, \boldsymbol{B}_1, \ldots, \boldsymbol{B}_m) \wedge G(\boldsymbol{B}_1, \ldots, \boldsymbol{B}_m)].$$

7. Applications to decidability and definability

The decidability results in this section were obtained in an even stronger form in [6] by use of the general theory of automata on infinite trees. We wish to show that for some of the results of [6], the simpler theory of special automata is sufficient. Also, the decision procedures involving special automata require fewer computational steps than the procedures of [6].

THEOREM 23. *There exists an effective procedure of deciding for a special Σ-automaton $\mathfrak{A} = \langle S, M, s_0, F\rangle$ whether $T(\mathfrak{A}) \neq \emptyset$. If $c(S) = n$, then this procedure requires n^4 computational steps.*

Proof. By forming the $\{\alpha\}$-table $M'(s, \alpha) = \bigcup_{\sigma \in \Sigma} M(s, \sigma)$, we pass to the case of a single letter alphabet $\{\alpha\}$. Clearly $T(\mathfrak{A}) \neq \emptyset$ if and only if $T(\langle S, M', s_0, F\rangle) \neq \emptyset$. Denote $M'(s, \alpha) = M(s)$. Since every tree has just one $\{\alpha\}$-valuation, we shall just talk about the $\{\alpha\}$-tree T or the finite $\{\alpha\}$-tree $E \subset T$. Thus assume $\mathfrak{A}$ to be an $\{\alpha\}$-automaton. Let $H \subseteq S$, denote by $R(H)$ the set of all $s \in S$ such that there exists a finite tree $E \neq \{\Lambda\}$ and an $\mathfrak{A}$-run $r: E \to S$ such that $r(\Lambda) = s$ and $r(Ft(E)) \subseteq H$.

An algorithm to compute $R(H)$ will proceed as follows. Put $H_0 = \emptyset$, and inductively for $i < \omega$,

$$H_{i+1} = H_i \cup \{s | \exists s_1 \exists s_2[(s_1, s_2) \in M(s), \{s_1, s_2\} \subseteq H \cup H_i]\}.$$

Now, $H_i \subseteq H_{i+1}$ for $i < \omega$, and if $H_i = H_{i+1}$ then $H_i = H_{i+k} = R(H)$, $k < \omega$. Since $H_i \subseteq S$, we certainly have $H_n = H_{n+1} = R(H)$. Given H_i, the calculation of H_{i+1} requires at most n^2 steps since $c(H_i) \leqq n$. Because $H_n = R(H)$, the calculation of $R(H)$ requires at most n^3 steps.

Define $F_0 = F$, and inductively for $i < \omega$, $F_{i+1} = F_i \cap R(F_i)$. Thus $F_{i+1} \subseteq F_i$ for $i < \omega$. Also, $F_i = F_{i+1}$ implies $F_i = F_{i+k}$ for $k < \omega$. Thus certainly $F_n = F_{n+1}$; put $F_n = G$. Note that the calculation of G requires at most $c(F) \leqq n$ calculations of sets $R(H)$, i.e. at most n^4 steps.

We claim that $T(\mathfrak{A}) \neq \emptyset$ if and only if $s_0 \in R(G)$. That $T \in T(\mathfrak{A})$ implies $s_0 \in R(G)$ is proved by methods similar to those of the proof of Theorem 27 and is left to the reader.

Assume $s_0 \in R(G)$ and let us construct an accepting $\mathfrak{A}$-run $r: T \to S$, thereby showing $T \in T(\mathfrak{A})$. Let $t_0 = (r_0, E_0)$ be a non-trivial finite S-tree such that $r_0 \in \mathrm{Rn}(\mathfrak{A}, E_0)$, $r_0(\Lambda) = s_0$, and $r_0(Ft(E_0)) \subseteq G$. For each $s \in G$ let $t_s = (r_s, E_s)$ be a non-trivial finite S-tree such that $r_s \in \mathrm{Rn}(\mathfrak{A}, E_s)$, $r_s(\Lambda) = s$, and $r_s(Ft(E_s)) \subseteq G$. Graft t_0 onto T at Λ; call the resulting (partial) S-tree (r_0, T). For each $x \in Ft(E_0)$, graft $t_{r_0(x)}$ onto (r_0, T) at x; call the resulting tree (r_1, T). Let $E_1 \subset T$ be the subtree of T on which r_1 is defined. We have $r_1(Ft(E_1)) \subseteq G$. For each $x \in Ft(E_1)$, graft $t_{r_1(x)}$ onto (r_1, T) at x; call the resulting tree—(r_2, T). Continuing in this manner, we get a sequence of partial S-trees (r_i, T), $i < \omega$ and finite trees $E_i = D(r_i)$, $i < \omega$. Because $E_s \neq \{\Lambda\}$ for $s \in G$, we have $Ft(E_i) < Ft(E_{i+1})$, $i < \omega$. Thus $T = \bigcup_{i<\omega} E_i$. Let $r = \lim_{i\to\infty} r_i$. Then $r: T \to S$ is an $\mathfrak{A}$-run on T, $r(\Lambda) = r_0(\Lambda) = s_0$, and $r(Ft(E_i)) = r_i(Ft(E_i)) \subseteq G \subseteq F$, $i < \omega$. Hence, $T \in T(\mathfrak{A}) \neq \emptyset$.

The following result follows from the decidability of S2S proved in [6]. Here we deduce it from the theory of special automata.

THEOREM 24. *There exists an effective procedure of deciding for every sentence G of the form $\exists A_1 \ldots \exists A_n F(A_1, \ldots, A_n)$* or *$\forall A_1 \ldots \forall A_n F(A_1, \ldots, A_n)$, where F is a formula with quantification over just individual of finite-set variables, whether $\mathcal{N}_2 \models G$.*

Proof. It is enough to consider the case of sentences G with existential quantifiers, If G is universal, then $\sim G$ is existential of the above form and $\mathcal{N}_2 \models G$ if and only if *not* $\mathcal{N}_2 \models \sim G$.

Assume that $G = \exists A_1 \ldots \exists A_n F(A_1, \ldots, A_n)$, where F is as above. To decide whether $\mathcal{N}_2 \models G$, construct by the procedure given in the proof of Theorem 19 a special $\{0,1\}^n$-automation $\mathfrak{A}$ which represents the relation $\{(A_1, \ldots, A_n) \mid \mathcal{N}_2 \models F(A_1, \ldots, A_n)\}$. Now $\mathcal{N}_2 \models G$ if and only if $T(\mathfrak{A}) \neq \emptyset$ and the latter question is decidable by Theorem 23.

The following theorem is due to Läuchli [4]. A stronger form is found in [6].

THEOREM 25. *The weak second-order theory WTO of linearly ordered sets is decidable.*

Proof. It is a well known result of Tarski that the downward Skolem-Löwenheim theorem holds for weak second-order languages. This implies that WTO is the same as the weak second-order theory of countable linearly ordered sets. Thus if G is a sentence of the weak second-order language of linear order, then $G \in WTO$ if and only if $P = \langle \bar{A}, \leqq \rangle \models G$ for every linearly ordered system P with $c(\bar{A}) \leqq \omega$.

In [6] we have shown that there exists a subset $B \subseteq T$ such that $\langle B, \preccurlyeq \rangle$

(see Lemma 20) has the order type of the rationals. This implies that for every countable linearly ordered system P there exists a subset $A \subseteq T$ such that $P \simeq \langle A, \preccurlyeq \rangle$.

Let G be a sentence as above. From the formula $G(A)$ of S2S by replacing all occurrences of $\boldsymbol{x} \leqq \boldsymbol{y}$ by $\boldsymbol{x} \preccurlyeq \boldsymbol{y}$, and relativizing all individual and finite-set quantifiers to $\boldsymbol{A}$. The formula $G(\boldsymbol{A})$ does not contain arbitrary set quantifiers. If $A \subseteq T$ then $\langle A, \preccurlyeq \rangle \models G$ if and only if $\mathscr{N}_2 \models G(A)$. Hence $G \in WTO$ if and only if $\mathscr{N}_2 \models \forall \boldsymbol{A} G(\boldsymbol{A})$. The last question is decidable by Theorem 24.

As in [6], the decision procedure given here is primitive recursive (in fact, even elementary recursive) which improves upon Läuchli's corresponding result [4].

The following theorem is also a special case of a result in [6]. It is an improvement of Ehrenfeucht's result [2] both in extending his result to weak second-order logic, and in yielding a primitive recursive decision procedure.

THEOREM 26. *The weak second-order theory WSU of a unary function is decidable.*

Proof. In the proof of Theorem 2.4 of [6], we constructed two formulas $F(\boldsymbol{x}, \boldsymbol{y}, \boldsymbol{C})$ and $Al(\boldsymbol{A}, \boldsymbol{C})$ with the following properties. For $x \in T$ and $C \subseteq T$ there exists at most one $y \in T$ such that $\mathscr{N}_2 \models F(x, y, C)$. Thus for a fixed $C \subseteq T$, $F(x, y, C)$ defines a binary relation $f \subseteq T \times T$ which is a mapping from $D(f) = \{x \mid \mathscr{N}_2 \models F(x, y, C)\}$ into T. For $A \subseteq T, C \subseteq T, \mathscr{N}_2 \models Al(A, C)$ holds if and only if $\langle A, f \mid A \rangle$ is an algebra, i.e. $f(A) \subseteq A$. For every countable unary algebra $P = \langle B, g \rangle$ there exist sets $A \subseteq T$, $C \subseteq T$ such that with the above notations, $\mathscr{N}_2 \models Al(A, C)$ and $P \simeq \langle A, f \mid A \rangle$. Examination of the construction of $F(\boldsymbol{x}, \boldsymbol{y}, \boldsymbol{C})$ and $Al(\boldsymbol{A}, \boldsymbol{C})$ will show that these formulas do not involve quantification over arbitrary set variables.

An argument similar to the one in the previous proof shows that for every sentence G of the weak second-order theory of a unary function there exists a sentence $\bar{G} = \forall \boldsymbol{A} \forall \boldsymbol{A} G(\boldsymbol{A}, \boldsymbol{C})$ of S2S, where G has no arbitrary set quantifiers, such that $G \in WSU$ if and only if $\mathscr{N}_2 \models \bar{G}$. This shows the decidability of WSU.

8. Characterization of weakly definable relations

We wish to show that a relation $R \subseteq P(T)^n$ is weakly-definable if and only if both R and its complement $P(T)^n - R$ are representable by special automata. To this end we study, for special automata, the question when is $T(\mathfrak{A}) \cap T(\mathfrak{B}) \neq \emptyset$.

Let $\mathfrak{A} = \langle S, M, s_0, F \rangle$, $\mathfrak{B} = \langle S', M', s_0', F' \rangle$. If $(v, T) = t \in T(\mathfrak{A}) \cap T(\mathfrak{B})$, then there are two accepting runs $r \in \mathrm{Rn}(\mathfrak{A}, t)$ and $r' \in \mathrm{Rn}(\mathfrak{B}, t)$. Hence there exists a finite subtree $E \subset T$ and two frontiers (of T) G and G' such that $G < Ft(E)$, $G' < Ft(E)$, $r(G) \subseteq F$ and $r'(G') \subseteq F'$. For every $x \in Ft(E)$ there exists a finite subtree $E_1 \subseteq T_x$, frontiers G_1 and G_1' (of T_x) such that $G_1 < Ft(E_1)$, $G_1' < Ft(E_1)$, $r(G_1) \subseteq F$, and $r'(G_1') \subseteq F'$. And so on, for the nodes $x \in Ft(E_1)$. These considerations motivate the following construction of a sequence of subsets of $S \times S'$.

Define $H_0 = S \times S'$. Define H_{i+1} inductively on i by: $(s, s') \in H_{i+1}$ if and only if $(s, s') \in H_i$ and there exists a finite Σ-tree $e = (v, E)$, where $E \neq \{\Lambda\}$, frontiers $G < Ft(E)$ and $G' < Ft(E)$, and runs $r \in \mathrm{Rn}(\mathfrak{A}, e)$, $r' \in \mathrm{Rn}(\mathfrak{B}, e)$ such that: (1) $r(\Lambda) = s$, $r'(\Lambda) = s'$; (2) $r(G) \subseteq F$, $r'(G') \subseteq F'$; (3) for every $x \in Ft(E)$ we have $(r(x), r'(x)) \in H_i$. Obviously, $H_0 \supseteq H_1 \supseteq \ldots$. Also, if $H_i = H_{i+1}$, then $H_i = H_{i+k}$ for every $k < \omega$. Thus, if $c(S)c(S') = m$, then certainly $H_m = H_{m+k}$ for $k < \omega$. With the above notations we have the following.

THEOREM 27. *$T(\mathfrak{A}) \cap T(\mathfrak{B}) \neq \emptyset$ if and only if $(s_0 s_0') \in H_m$.*

Proof. Assume $(v, T) = t \in T(\mathfrak{A}) \cap T(\mathfrak{B})$. There exist runs $r \in \mathrm{Rn}(\mathfrak{A}, t)$ and $r' \in \mathrm{Rn}(\mathfrak{B}, t)$ such that $r(\Lambda) = s_0$, $r'(\Lambda) = s_0'$ and for every path $\pi \subset T$, $\mathrm{In}(r \mid \pi) \cap F \neq \emptyset$, $\mathrm{In}(r' \mid \pi) \cap F' \neq \emptyset$. This implies the existence of a strictly increasing sequence $(E_i)_{i \leqq m}$ of finite subtrees $E_i \subset E_{i+1} \subset T$, $i < m$, such that $E_0 = \{\Lambda\}$ and, for each $i < m$, there are two frontiers (of T) G_i, G_i' satisfying $Ft(E_i) \leqq G_i < Ft(E_{i+1})$, $r(G_i) \subseteq F$, $Ft(E_i) \leqq G_i' < Ft(E_{i+1})$, $r'(G_i) \subseteq F'$.

Let $x \in Ft(E_i)$, $i \leqq m-1$ and consider the finite tree (with root x) $E = T_x \cap E_{i+1}$. Note that $Ft(E) = T_x \cap Ft(E_{i+1})$ and that $G_i \cap T_x$ and $G_i' \cap T_x$ are frontiers of E (and of T_x). Now, $r \mid E$ and $r' \mid E$ are, respectively, $\mathfrak{A}$-and $\mathfrak{B}$-runs on (v, E). Also $r(G_i \cap T_x) \subseteq F$ and $r'(G_i \cap T_x) \subseteq F'$. Hence $(r(x), r'(x)) \in H_1$.

Assume now that $i \leqq m-2$, then $i+1 \leqq m-1$ and by the previous result $(r(y), r'(y)) \in H_1$ for every $y \in Ft(E_{i+1})$. Thus, by considering the same tree (v, E), we have $(r(x), r'(x)) \in H_2$.

Using induction, we get $(r(x), r'(x)) \in H_k$ for $x \in Ft(E)$, $i \leqq m-k$. Since $\Lambda \in Ft(E_0)$ we have, in particular, $(s_0, s_0') = (r(\Lambda), r'(\Lambda)) \in H_m$.

To prove the converse assertion, let $(s_0, s_0') \in H_m$ Because $H_m = H_{m+1}$, we have for every $(s, s') \in H_m$ a finite Σ-tree $e(s, s') = (v(s, s'), E(s, s'))$ where $E(s, s') \neq \{\Lambda\}$, runs $r(s, s') = r \in \mathrm{Rn}(\mathfrak{A}, e(s, s'))$, $r'(s, s') = r' \in \mathrm{Rn}(\mathfrak{B}, e(s, s'))$, and frontiers $G < Ft(E(s, s'))$, $G' < Ft(E(s, s'))$ satisfying (1) $r(\Lambda) = s$, $r'(\Lambda) = s'$; (2) $r(G) \subseteq F$, $r'(G') \subseteq F'$; (3) $x \in Ft(E(s, s'))$ implies $(r(x), r'(x)) \in H_m$.

The notion of grafting a subtree t onto a tree (v, T) at $x \in T$, can be extended to the case that $t = (u, E)$ is a finite tree and the valuation is not completely defined. Namely, the result of this graft is defined to be the tree (w, T) such that $w(y) = v(y)$ if $y \notin xE$ and $w(xz) = u(z)$ for $z \in E$.

Let t_1 be the result of grafting $e(s_0, s_0')$ onto $(\emptyset, T)$ at Λ. On $t_1 = (w_1, T)$ define a partial $\mathfrak{A}$-run r_1 by $r_1(x) = r(s, s_0')(x)$ for $x \in E(s_0, s_0')$, and a partial $\mathfrak{B}$-run r_1' by $r_1'(x) = r'(s_0, s_0')(x)$, $x \in E(s_0, s_0')$. Note that for $x \in Ft(E(s_0, s_0'))$, $(r_1(x), r_1'(x)) \in H_m$.

Denote $E(s_0, s_0') = E_1$. Graft simultaneously onto t_1, at each $x \in Ft(E_1)$, the tree $e(r_1(x), r_1'(x))$ and call the resulting tree $t_2 = (w_2, T)$. The domain of definition of w_2 is a finite tree, call it E_2. Extend r_1 and r_1 to runs $r_2 \in \mathrm{Rn}(\mathfrak{A}, (w_2, E_2))$, $r_2' \in \mathrm{Rn}(\mathfrak{B}, (w_2, E_2))$ as follows. If $y \in (E_2 - E_1) \cup Ft(E_1)$ then there is a unique $x \in Ft(E_1)$ such that $y \in xE(s, s')$, where $r_1(x) = s$, and $r_1'(x) = s'$. Assume $y = xz$, $z \in E(s, s')$ and set $r_2(y) = r(s, s')(z)$, $r_2'(y) = r'(s, s')(z)$. Note that for $y \in Ft(E_1)$ we have $r_1(y) = r_2(y)$, $r_1'(y) = r_2'(y)$. Thus r_2 and r_2' indeed extend r_1 and r_1', respectively. Also, for $x \in Ft(E_2)$ we have $(r_2(x), r_2'(x)) \in H_m$. Thus the process can be continued by induction to obtain sequences of partial Σ-trees $(t_i = (w_i, T))_{i<\omega}$, of finite trees $((w, E_i))_{i<\omega}$, and of runs $(r_i)_{i<\omega}$ and $(r_i)_{i<\omega}$. Let $t = \lim_{i<\omega} t_i$; $r = \lim_{i<\omega} r_i$; and $r' = \lim_{i<\omega} r_i'$. Then, $r \in \mathrm{Rn}(\mathfrak{A}, t)$, $r' \in \mathrm{Rn}(\mathfrak{B}, t)$, $r(\Lambda) = s_0'$, $r'(\Lambda) = s_0'$. Our construction implies that for evevery $x \in Ft(E_i)$, $r \mid (T_x \cap E_{i+1})$ and $r' \mid (T_x \cap E_{i+1})$ coincide with some $r(s, s')$ and $r'(s, s')$, respectively. This entails the existence of two frontiers, G_i, G_i' of T, $Ft(E_i) \leqq G_i < Ft(E_{i+1})$ $Ft(E_i) \leqq G_i' < Ft(E_{i+1})$, such that $r(G_i) \subseteq F$, $r'(G_i) \subseteq F'$. Thus r is an accepting $\mathfrak{A}$-run for t and r' is an accepting $\mathfrak{B}$-run for t. Hence $t \in T(\mathfrak{A}) \cap T(\mathfrak{B})$.

Keeping the notations of the previour theorem we have

COROLLARY 28. *If there exists a Σ-tree $t = (v, T)$ and a strictly increasing sequence $(E_i)_{i \leqq m}$ of finite trees as in the first part of the previous proof, then $T(\mathfrak{A}) \cap T(\mathfrak{B}) \neq \emptyset$.*

Proof. It was shown before that our assumption implies $(s_0, s_0') \in H_m$. Hence $T(\mathfrak{A}) \cap T(\mathfrak{B}) \neq \emptyset$.

To simplify notations we shall now restrict ourselves to Σ^n $(= \{0,1\}^n)$-trees. We shall talk interchangeably about a sequence $A = (A_1, \ldots, A_n) \in P(T)^n$ and the corresponding Σ^n-tree $\tau(A)$.

Let $t = (v, T)$ be a tree accepted by $\mathfrak{A}$. We wish to approximate this fact by certain statements which are weakly definable. On t there exists a run $r \in \mathrm{Rn}(\mathfrak{A}, t)$ and an infinite strictly increasing sequence of finite subtrees $(G_i)_{i<\omega}$ such that $r(\Lambda) = s_0$, $r(Ft(G_i)) \subseteq F$, $i < \omega$. This implies that for every finite subtree $E \subset T$ there exists a subtree $E \subset G \subset T$ and a run $r' \in \mathrm{Rn}(\mathfrak{A}, (v, G))$ such that $r'(\Lambda) = s_0$, $r'(Ft(G)) \subseteq F$. Namely, $G = G_i$

for an appropriate i, and $r' = r|G$. This r' has the property that for every $x \in Ft(G)$ and every finite subtree $E' \subset T_x$ with root x, there exists a finite tree $E' \subset G' \subset T_x$ and a run $r'' \in \mathrm{Rn}(\mathfrak{A}, (v, G'))$ such that $r''(x) = r'(x)$, $r''(Ft(G')) \subseteq F$. Namely, $G' = G_j \cap T_x$ for an appropriate $j > i$, and $r'' = r|G'$. And so on. These facts will now be formalized by an inductive definition.

Define $K_0 = S \times T \times V_{\Sigma,\Lambda}$, i.e. $K_0(s,x,t)$ is always true. Let $s \in S$, $x \in T$, (v, T) be a Σ^n-tree, and $i < \omega$; define

(6) $K_{i+1}(s,x,(v,T)) = \forall E \exists G \exists r[E$ *finite subtree of* $T_x \rightarrow E \subseteq G$ *is a finite subtree* of $T_x \wedge r \in \mathrm{Rn}(\mathfrak{A}, (v, G)) \wedge r(x) = s \wedge r(Ft(G)) \subseteq F \wedge \forall y[y \in Ft(G) \rightarrow K_i(r(y), y, (v, T))]]$.

For every fixed $i < \omega$ and fixed $s \in S$, K_i is a relation $K_i \subseteq T \times P(T)^n$. This relation is weakly definable in $\mathcal{N}_2$. Namely, let $\eta: S \rightarrow \Sigma^k$ be a fixed coding.

In (6), the finite trees G, E, are represented by finite sets α, β. The Σ^n-tree (v, T) is, of course, represented by an n-tuple $(A_1, \ldots, A_n)$ of arbitrary sets. These set variables remain free. We recall that the relation $x \leqq y$ is weakly definable. Hence the relation (between $G \subset T$ and $x \in T$): G is a finite subtree of T_x with root x, is weakly definable. Now, a run $r: G \rightarrow S$ is represented, via η, by k finite sets $\alpha_1, \ldots, \alpha_k$. A relation $r(x) = s$ is then weakly definable. Finally, $K_i(r(y), y, t)$ is replaced by $\bigwedge_{s \in S}[r(y) = s \rightarrow K_i(s, y, t)]$.

THEOREM 29. *If $R \subseteq P(T)^n$ and $Q = P(T)^n - R$ are representable by special automata, then R is weakly-definable. Conversely, if R is weakly definable, then R and Q are representable by special automata.* (2)

Proof. The latter part of the theorem is simply Theorem 19 and requires no proof.

Let R be represented by $\mathfrak{A} = \langle S, M, s_0, F \rangle$, and Q be represented by $\mathfrak{B} = \langle S', M', s_0', F' \rangle$; let $c(S)c(S') = m$. Thus $T(\mathfrak{A}) \cap T(\mathfrak{B}) = \emptyset$. We claim that $t = (v, T) \in T(\mathfrak{A})$ if and only if $K_m(s_0, \Lambda, t)$ is true. Since this last relation is weakly definable, this will prove our theorem.

If $t \in T(\mathfrak{A})$ then $K_i(r(x), x, t)$ is true for every $i < \omega$, $x \in T$, by the remarks preceding the definition of $K_i(s, x, t)$. Hence $K_m(s_0, \Lambda, t)$ is true.

Assume by way of contradiction that $K_m(s_0, \Lambda, t)$ for $t = (v, T)$ is true but $t \notin T(\mathfrak{A})$, i.e. $t \in T(\mathfrak{B})$. Let $r' \in \mathrm{Rn}(\mathfrak{B}, t)$ be an accepting run of $\mathfrak{B}$ on t. Define $E_0 = \{\Lambda\}$ and let G_0' be a frontier of T such that $r'(G_0') \subseteq F'$. Since $K_m(s_0, \Lambda, t)$ is true, there exists a finite tree $\bar{G}_0$ and an $\mathfrak{A}$-run $r_0: \bar{G}_0 \rightarrow S$ on $(v, \bar{G}_0)$ such that $r_0(\Lambda) = s_0$ and, for $G_0 = Ft(\bar{G}_0)$, (1) $r_0 G(_0) \subseteq F$ and (2) $K_{m-1}(r_0(x), x, t)$ is true for all $x \in G_0$. Let $E_1 \subset T$ be a finite tree such

(2) This theorem was suggested by A. Tarski.

that $G_0, G'_0 < Ft(E_1)$. There exists a frontier G'_1 of T such that $Ft(E_1) \leqq G'_1$ and $r'(G'_1) \subseteq F'$. Applying the above statement (2) to each $x \in G_0$ and the finite subtree $T_x \cap E_1$ of T_x, we get the existence of a finite tree $E_1 \subseteq \bar{G}_1$ and an extension $r_1 \colon \bar{G}_1 \to S$ of r_0 such that r_1 is an $\mathfrak{A}$-run on $(v, \bar{G}_1)$ and, for $G_1 = Ft(\bar{G}_1)$, (1) $r_1(G_1) \subseteq F$ and (2) $K_{m-2}(r_0(x), x, t)$ is true for all $x \in G_1$. Continue in this way up to a tree E_{m-1} and two frontiers $Ft(E_{m-1}) \leqq G_{m-1}$, G'_{m-1}. Extend the run $r_{m-1} \colon \bar{G}_{m-1} \to S$ in some way to a run $r \in \mathrm{Rn}(\mathfrak{A}, t)$. Then $(E_i)_{i<m}$, r and r' satisfy the conditions of Corollary 28. Hence $T(\mathfrak{A}) \cap T(\mathfrak{B}) \neq \emptyset$, a contradiction. Thus $K_m(s_0, \Lambda, t)$ is true if and only if $t \in T(\mathfrak{A})$ and the proof is completed.

THEOREM 30. *A formula* $H(\boldsymbol{A}_1, \ldots, \boldsymbol{A}_n)$ *of* S2S *is equivalent in* S2S *with some formula* $H_1(\boldsymbol{A}_1, \ldots, \boldsymbol{A}_n)$ *which has no arbitrary set quantifiers, if and only if there exist two formulas* $G_1 = \forall \boldsymbol{B}_1 \ldots \forall \boldsymbol{B}_p F_1(\boldsymbol{A}_1, \ldots, \boldsymbol{A}_n, \boldsymbol{B}_1, \ldots \boldsymbol{B}_p)$ *and* $G_2 = \exists \boldsymbol{B}_1 \ldots \exists \boldsymbol{B}_q F_2(\boldsymbol{A}_1, \ldots, \boldsymbol{A}_n, \boldsymbol{B}_1, \ldots, \boldsymbol{B}_q)$, *such that* F_i *has no arbitrary set quantifier and* H *is equivalent with* G_i *in* S2S *for* $i = 1, 2$.

Proof. This follows at once from Theorems 22 and 29.

We conclude our discussion with a decision problem suggested by H. Gaifman. Is it decidable to determine for every given formula $H(\boldsymbol{A}_1, \ldots, \boldsymbol{A}_n)$ of S2S whether there exists a formula $H_1(\boldsymbol{A}_1, \ldots, \boldsymbol{A}_n)$ without arbitrary set quantifiers which is equivalent to H in S2S?

A positive solution for this problem will have interesting consequences. We have seen in §7 that every countable linearly ordered set $P = \langle \bar{A}, \leqq \rangle$ is reproducible (up to isomorphism) as $\langle A, \preccurlyeq \rangle$ where $A \subseteq T$. Thus every question concerning weak definability of a relation definable in the (monadic) second-order theory of countable linear order can be converted into a corresponding question for S2S. If the latter question can be effectively answered so can the original question. Similar remarks apply to the second-order theory of a unary function and to other theories.

REFERENCES

[1] J. E. DONER, *Decidability of the weak second-order theory of two successors*, Notices Amer. Math. Soc., **12** (1965), 819.

[2] A. EHRENFEUCHT, *Decidability of the theory of one function*, Notices Amer. Math. Soc., **6** (1959), 268.

[3] D. M. GABBAY, *Decidability results in non-classical logics I*, Technical Report No. 29, Applied Logic Branch, Jerusalem, 1969, p. 42.

[4] H. LÄUCHLI, *A decision procedure for the weak second order theory of linear order, Contributions to mathematical logic*, K. Schütte, editor, North-Holland, Amsterdam, 1968, 189–197.

[5] M. O. RABIN, *Mathematical theory of automata*, Proc. Sympos. Appl. Math., Vol. 19, Amer. Math. Soc., Providence, R.I., 1968, 153–175.

[6] ———, *Decidability of second order theories and automata on infinite trees*, Trans. Amer. Math. Soc., **141** (1969), 1–35.

[7] J. W. THATCHER and J. B. WRIGHT, *Generalized finite automata*, Notices Amer. Math. Soc., **12** (1965), 820.

DETERMINACY AND PREWELLORDERINGS OF THE CONTINUUM

YIANNIS N. MOSCHOVAKIS†
University of California, Department of Mathematics, Los Angeles, California, U.S.A.

Let Λ be a class of subsets of $R \times R$, where $R = {}^{\omega}\omega$ = the set of number theoretic functions. An ordinal ξ is *realized* in Λ if there is a relation in Λ which is a *prewellordering* of length ξ (the precise definition of prewellordering is given in §3). In this paper we study

$$o(\Lambda) = \mathit{supremum}\{\xi : \xi \text{ is realized in } \Lambda\}$$

for various Λ's, in the context of a set theory that assumes full determinacy, *AD* (most of the time), and dependent choices for sets of reals, *DC* (some of the time). We are particularly interested in the projective classes $\boldsymbol{\Delta}_k^1$ and our most quotable theorems concern these: *If AD, then for each* $k \geqq 1$, $o(\boldsymbol{\Delta}_k^1)$ *is a cardinal. If AD and DC, then for each* $k \geqq 1$, $o(\boldsymbol{\Delta}_k^1) \geqq \aleph_k^1$; *also for each odd* $k \geqq 1$, $o(\boldsymbol{\Delta}_k^1)$ *is a regular cardinal and each subset of* $o(\boldsymbol{\Delta}_k^1)$ *is "*$\boldsymbol{\Pi}_k^1$ *in the codes", in a reasonable sense of this expression.* However, our results apply to interesting larger classes of definable sets, e.g. $\boldsymbol{\Delta}_k^2$, and even to the full power of $R \times R$.

In §§1, 2 we establish notation and give a brief summary of the properties of recursive functions, projective sets and determinacy that we need. The main new constructions of this paper are in §3, the consequences when $\Lambda = R \times R$ in §4, the consequences when Λ is assumed to satisfy various structure properties in §5. In §§6,7 we collect the results about the projective classes and the classes $\boldsymbol{\Delta}_k^n$, respectively.

I wish to thank R. M. Solovay for many discussions during the Jerusalem meeting which led to a substantial improvement of the results reported here.

The problems of this paper were very much in my mind in the period immediately after my father's death, when his loss was felt very keenly; the final manuscript was completed in his beloved Phaliron house one year after his death, to the day. For these, and for many other reasons, I wish to dedicate this paper to my father's memory.

---

† The author is a Guggenheim fellow. Work on this paper was sponsored in part by an N.S.F. grant.

§1. Preliminaries. Let $\omega = \{0, 1, 2, \ldots\}$ be the set of natural numbers $R = {}^{\omega}\omega$ the set of number-theoretic functions or (for our purposes) *real numbers*. We study subsets of all product spaces

$$\mathfrak{X} = X_1 \times \ldots \times X_k$$

where each X_i is ω or R; we use x, y, z, w as variables over such product spaces, $k, l, m, n, t, s, \ldots$ as variables over ω and $\alpha, \beta, \gamma, \ldots$ as variables over R. If x is an i-tuple and y a j-tuple, we let (x, y) be the $i+j$-tuple whose first i components are the components of x and whose last j components are the components of y. Similarly, if $\mathfrak{X} = X_1 \times \cdots \times X_i$ and $\mathfrak{Y} = Y_1 \times \cdots \times Y_j$, then

$$\mathfrak{X} \times \mathfrak{Y} = X_1 \times \ldots \times X_i \times Y_1 \times \ldots \times Y_j.$$

It will be necessary to assume a minimum knowledge of recursion theory which can be easily acquired from Kleene 1952 (or any elementary text), in particular the so-called *enumeration*, *iteration* (S^m_n) and (second) *recursion* theorems. The approach to recursive and continuous partial functions on $\mathfrak{X}$ to $\mathfrak{Y}$ for product spaces $\mathfrak{X}$, $\mathfrak{Y}$ that we outline here is substantially that of Kleene 1959.

As usual,

$$\langle n_0, \ldots, n_k \rangle = 2^{n_0+1} \cdot 3^{n_1+1} \cdot \ldots \cdot p_k^{n_k+1},$$

with p_i the ith prime,

$$(n)_i = \textit{supremum}\,\{j : p_i^{j+1} \textit{ divides } n\},$$

$$(\alpha)_i = \lambda t \alpha(\langle i, t \rangle).$$

Put

$$\bar{\alpha}(0) = 1, \quad \bar{\alpha}(t+1) = \langle \alpha(0), \ldots, \alpha(t) \rangle,$$

$$\bar{n}(t) = n,$$

and if $x = (x_1, \ldots, x_k)$ with $x_i \in \omega$ or $x_i \in R$,

$$\bar{x}(t) = (\bar{x}_1(t), \ldots, \bar{x}_k(t)) \in \omega^k.$$

A partial function $f : \mathfrak{X} \to \omega$ is recursive if there exist recursive partial functions $g(n_1, \ldots, n_k)$, $h(n_1, \ldots, n_k)$ such that

$$f(x) \simeq g(\bar{x}(s)), \; s \simeq \mu t[(\forall t' < t)[h(\bar{x}(t')) \simeq 0] \;\&\; h(\bar{x}(t)) > 0].$$

Clearly $f : \mathfrak{X} \to \omega$ is recursive in this sense if $f(x)$ depends only on a finite initial segment of the real components of x and in an effective way.

A partial function $f : \mathfrak{X} \to R$ is recursive if there exists a recursive partial function $g : \mathfrak{X} \times \omega \to \omega$ such that

$$f(x) \simeq \lambda t g(x,t);$$

in particular

$$Domain(f) = \{x : \forall t[g(x,t) \text{ is defined}]\}.$$

Finally, a partial function $f : \mathfrak{X} \to \mathfrak{Y} = Y_1 \times \ldots \times Y_l$ is recursive if there exist recursive partial functions $f_1, \ldots, f_l$ on $\mathfrak{Y}$ to ω or R as required so that

$$f(x) = (f_1(x), \ldots, f_l(x)).$$

We relativize this concept by introducing parameters: a partial function $f : \mathfrak{X} \to \mathfrak{Y}$ is recursive in β (or β-recursive) if there is a recursive $g : R \times \mathfrak{X} \to \mathfrak{Y}$ such that

$$f(x) \simeq g(\beta, x).$$

Each product space $\mathfrak{X}$ is a topological space, in fact a metric space, where we take ω as discrete and then ${}^{\omega}\omega$ and each $\mathfrak{X}$ with the product topologies. It is easy to verify that if $f : \mathfrak{X} \to \mathfrak{Y}$ is β-recursive and totally defined, then it is continuous. This suggests an extension of the continuity property to partial functions: a partial $f : \mathfrak{X} \to \mathfrak{Y}$ is (partial) *continuous* if it is β-recursive for some β. Although the topological significance of continuous partial functions is not immediately apparent, they are a very useful class because a decent part of recursion theory can be generalized to them.

Let $type(\mathfrak{X}) = 0$ if $\mathfrak{X}$ is a product of copies of ω, $type(\mathfrak{X}) = 1$ if at least one factor of $\mathfrak{X}$ is R. One easily verifies that if $type(\mathfrak{X}) = type(\mathfrak{Y})$, then there exists a recursive function

$$\pi : \mathfrak{X} \to \mathfrak{Y}$$

which has a recursive inverse and is a topological homeomorphism. These canonical homeomorphisms give us in particular recursive pairing functions, i.e. recursive bijections on $\mathfrak{X} \times \mathfrak{X}$ to $\mathfrak{X}$ with recursive inverses.

It is not hard to prove the following characterization of continuous partial functions: If $type(\mathfrak{Y}) = 0$, *then a partial function* $f : \mathfrak{X} \to \mathfrak{Y}$ *is continuous if and only if* $Domain(f)$ *is open and* f *is continuous on its domain. If* $type(\mathfrak{Y}) = 1$, *then a partial* $f : \mathfrak{X} \to \mathfrak{Y}$ *is continuous if and only if* $Domain(f)$ *is a* G_δ *set (a countable intersection of open sets) and* f *is continuous on its domain.*

The following three basic results about continuous partial functions can be deduced easily from the corresponding results about recursive functions on ω to ω and the definitions above.

PARAMETRIZATION THEOREM (*the analog of enumeration*). *For each* $\mathfrak{X}$, $\mathfrak{Y}$ *there is a recursive partial function* $\Phi : R \times \mathfrak{X} \to \mathfrak{Y}$ *such that each continuous partial* $f : \mathfrak{X} \to \mathfrak{Y}$ *is given by*

$$f(x) \simeq \Phi(\varepsilon, x)$$

for some $\varepsilon \in R$. To conform with standard notation of recursion theory we put

$$\Phi(\varepsilon, x) \simeq \{\varepsilon\}(x) \simeq \{\varepsilon\}^{\mathfrak{X},\mathfrak{Y}}(x),$$

where we shall put the superscripts in only when necessary to avoid confusion.

ITERATION THEOREM. *For each* $\mathfrak{X}$ *there is a recursive function* $S^{\mathfrak{X}}: R \to R$ *such that for all* $\mathfrak{Y}, \mathfrak{Z}$

$$\{\varepsilon\}(x, y) \simeq \{S^{\mathfrak{X}}(\varepsilon, x)\}(y),$$

i.e. with the superscripts

$$\{\varepsilon\}^{\mathfrak{X}\times\mathfrak{Y},\mathfrak{Z}}(x, y) \simeq \{S^{\mathfrak{X}}(\varepsilon, x)\}^{\mathfrak{Y},\mathfrak{Z}}(y).$$

RECURSION THEOREM. *For each* $\mathfrak{X}$ *there is a recursive function* $F^{\mathfrak{X}}: R \to R$ *such that if* $\varepsilon^* = F^{\mathfrak{X}}(\varepsilon)$, *then for every* $\mathfrak{Z}$

$$\{\varepsilon\}(\varepsilon^*, x) \simeq \{\varepsilon^*\}(x).$$

We shall be studying classes Λ, Γ, Δ of subsets of R. For any $\mathfrak{X}$, any such class Λ put

(1–1) $[\mathfrak{X}]\Lambda = \{A \subseteq \mathfrak{X}$: *for some* $B \in \Lambda$ *and some recursive* $f: \mathfrak{X} \to R$, $A = f^{-1}[B]\}$.

The notation $[\mathfrak{X}]\Lambda$ is cumbersome and we shall avoid it by writing "$A \in \Lambda$" or even "A is Λ" for $A \subseteq \mathfrak{X}$ when it is clear (or irrelevant) which space $\mathfrak{X}$ is involved. There is a confusion in this convention if $\mathfrak{X} = R$ unless Λ is closed under recursive preimages, but all the classes we care about will satisfy this condition.

The following operations on classes of sets will be useful:

$\&\Lambda = \{A \cap B\colon A, B \in \Lambda\}$ (*conjunction*, &)

$\bigvee\Lambda = \{A \cup B\colon A, B \in \Lambda\}$ (*disjunction*, $\bigvee$)

$\neg\Lambda = \{R - A\colon A \in \Lambda\}$ (*negation* or *dual*, $\neg$)

$\Sigma^0\Lambda = \{A$: *for some* $B \in \Lambda$, $B \subseteq \omega \times R$, $\alpha \in A \Leftrightarrow (\exists n)[(n, \alpha) \in B]\}$ (*existential number quantification*, $\exists m$)

$\Sigma^1\Lambda = \{A$: *for some* $B \in \Lambda$, $B \subseteq R \times R$, $\alpha \in A \Leftrightarrow (\exists\beta)[(\beta, \alpha) \in B]\}$ (*existential real quantification*, $\exists\alpha$)

$\Pi^0\Lambda = \neg\Sigma^0\neg\Lambda$ (*universal number quantification*, $\forall m$)

$\Pi^1\Lambda = \neg\Sigma^1\neg\Lambda$ (*universal real quantification*, $\forall\alpha$).

The classical projective classes are defined starting with

$$\boldsymbol{\Sigma}^1_0 = \{A \subseteq R : A \textit{ is open}\}$$

and then proceeding inductively,

$$\begin{aligned} \boldsymbol{\Pi}^1_k &= \neg\boldsymbol{\Sigma}^1_k \, , \\ \boldsymbol{\Sigma}^1_{k+1} &= \Sigma^1\boldsymbol{\Pi}^1_k \, , \\ \boldsymbol{\Delta}^1_k &= \boldsymbol{\Sigma}^1_k \bigcap \boldsymbol{\Pi}^1_k \, . \end{aligned}$$

Now A is open if and only if there is a continuous partial $f: R \to \omega$ such that $A = f^{-1}[\{0\}]$. We obtain so-called *light-face* classes if we start with preimages of $\{0\}$ by *recursive* partial functions,

$$\Sigma^1_0 = \{A \subseteq R\text{: } \textit{for some recursive partial } f: R \to \omega,\ \alpha \in A \Leftrightarrow f(\alpha) \cong 0\}$$

and then proceed inductively with the operators Σ^1, $\neg$, $\cap$ as before.

It is well-known that the projective classes are closed under some of the operations above, e.g. $\boldsymbol{\Sigma}^1_k$ is closed under continuous substitution (i.e. $B \in \boldsymbol{\Sigma}^1_k$ and $f: R \to R$ continuous $\Rightarrow f^{-1}[B] \in \boldsymbol{\Sigma}^1_k$), &, $\vee$, Σ^0, Σ^1 and for $k > 0$, Π^0. We shall use these closure properties extensively, sometimes without explicit mention.

A different approach to the projective classes is through definability in the language of *second order number theory* or *analysis*. The language of number theory has the usual logical symbols, number variables $n, m, \ldots$, $=$, the individual constant 0 and the function symbols $'$, $+$, $\cdot$. For analysis we add real variables $\alpha, \beta, \ldots$, so that e.g. $\alpha(n)$, $\beta(\alpha(n))$ are number-terms, and the corresponding quantifiers $\exists\alpha, \forall\alpha$. By a classical lemma of Gödel every recursive relation on ω is definable by a formula of number theory. From this it follows trivially that every recursive relation on R is definable by a formula in the language of analysis with no bound real variables. This in turn gives simple syntactical characterizations for $\boldsymbol{\Sigma}^1_k, \boldsymbol{\Pi}^1_k, \Sigma^1_k, \Pi^1_k$ when $k > 0$, e.g. a set $A \subseteq R$ is in $\boldsymbol{\Sigma}^1_3$ if and only if it is definable by a formula of the form

$$\exists\alpha_1 \forall\alpha_2 \exists\alpha_3 \theta(\beta_0, \alpha, \alpha_1, \alpha_2, \alpha_3)$$

where $\theta(\beta, \alpha, \alpha_1, \alpha_2, \alpha_3)$ has no real quantifiers and β_0 is a fixed real parameter.

One may study classes of sets which are definable in languages richer than second order number theory. Here we shall go only one step further, i.e. consider *third order number theory* which is obtained by adding variables $F, G, H, \ldots$ (to second order number theory) which range over ${}^R\omega$, so that $F(\alpha)$, $\beta(G(\alpha))$ are number-terms, and the corresponding quantifiers $\exists F$, $\forall F$. A subset A of R is $\mathbf{\Sigma}_1^2$ if it is definable by a formula of the form

$$\exists F\, \theta(\beta_0, \alpha, F)$$

where $\theta(\beta, \alpha, F)$ has no bound type-3 variables and β_0 is a fixed real parameter. (If we do not allow the parameter β_0, we obtain the class Σ_1^2.) As usual, $\mathbf{\Pi}_1^2 = \neg\mathbf{\Sigma}_1^2, \mathbf{\Delta}_1^2 = \mathbf{\Sigma}_1^2 \cap \mathbf{\Pi}_1^2$. The classes $\mathbf{\Sigma}_k^2, \mathbf{\Pi}_k^2$ $(k > 1)$ are defined by the obvious extension, e.g. for $\mathbf{\Sigma}_2^2$ we take formulas of the form

$$\exists F\, \forall G\, \theta(\beta_0, \alpha, F, G).$$

§2. **Structure properties.** We are interested in structure properties of classes of subsets of R or a fixed product space $\mathfrak{X}$. In this section we summarize a few fundamental such properties with which we shall be concerned and we outline briefly the known results about them. We let Λ, Γ, Δ stand for arbitrary (non-empty) classes of subsets of R and we recall the definition of $[\mathfrak{X}]\Lambda$ in (1–1).

2.1. *Universal sets.* A set $G \subseteq R \times \mathfrak{X}$ is (*R*-)*universal* for $[\mathfrak{X}]\Gamma$ if $G \in \Gamma$ and

$$[\mathfrak{X}]\Gamma = \{G_\alpha : \alpha \in R\},$$

where G_α is the *α-section* of G,

$$G_\alpha = \{x : (\alpha, x) \in G\}.$$

If $[\mathfrak{X}]\Gamma$ has G as a universal set, we say that $[\mathfrak{X}]\Gamma$ is (*R*-)*parametrized* (by G).

Suppose Γ is closed under recursive substitution (i.e. recursive preimages) and is parametrized by some $G \subseteq R \times R$. For each $\mathfrak{X}$, put

$$h(x) = \lambda t \langle \bar{x}(t) \rangle;$$

now $h : \mathfrak{X} \to R$ is an injection and it is easy to construct a surjection $g : R \twoheadrightarrow \mathfrak{X}$ which agrees with h^{-1} on the image $h[\mathfrak{X}]$. In particular, for each $B \subseteq \mathfrak{X}$, $h^{-1}g^{-1}[B] = B$. From this it follows that each $B \in [\mathfrak{X}]\Gamma$ is of the form $h^{-1}[A]$ for some $A \in \Gamma$ (where $A = g^{-1}f^{-1}[C]$, if $B = f^{-1}[C]$, with recursive f), and hence

$$\text{(2–1)} \qquad H = \{(\alpha, x) : (\alpha, hx) \in G\}$$

parametrizes $[\mathfrak{X}]\Gamma$. One then proceeds to show easily that for such Γ, $\neg\Gamma$, $\Sigma^0\Gamma$, $\Sigma^1\Gamma$, $\Pi^0\Gamma$, $\Pi^1\Gamma$ are also parametrized.

It is a classical result that the projective classes $\mathbf{\Sigma}_k^1$, $\mathbf{\Pi}_k^1$ are parametrized.

We can prove it here by taking for $\mathbf{\Sigma}_0^1$ the set

$$G = \{(\varepsilon, \alpha) \colon \{\varepsilon\}(\alpha) \simeq 0\}$$

and then proceeding inductively by the procedures above.

A simple diagonal argument shows that if Γ is closed under recursive substitution and G parametrizes Γ, then $G \notin \neg\Gamma$. Thus for each k, $\mathbf{\Sigma}_k^1 - \mathbf{\Pi}_k^1 \neq \emptyset$, and we obtain easily the *hierarchy theorem*,

$$\mathbf{\Delta}_0^1 \subsetneq \mathbf{\Delta}_1^1 \subsetneq \mathbf{\Delta}_2^1 \subsetneq \dots .$$

2.2 *Separation and reduction.* Γ satisfies the *separation property* if for each $A, B \in \Gamma$ with $A \cap B = \emptyset$, there exists $C \in \Gamma \cap \neg\Gamma$ with $A \subseteq C$, $B \cap C = \emptyset$. Γ satisfies the *reduction property* if for each $A, B \in \Gamma$ there exist $A_1, B_1 \in \Gamma$ with $A_1 \subseteq A, B_1 \subseteq B, A_1 \cap B_1 = \emptyset$ and $A_1 \cup B_1 = A \cup B$. One easily verifies that *Reduction*$(\Gamma) \Rightarrow$ *Separation*$(\neg\Gamma)$ — to separate two sets reduce their complements. A slightly trickier argument shows that if Γ is parametrized and closed under recursive substitution, then *Reduction*$(\Gamma) \Rightarrow$ *not Separation* (Γ). Thus under these hypotheses reduction cannot hold for both Γ and $\neg\Gamma$ and if it holds for one of these classes then separation holds for the other.

It has been known from the classical work in descriptive set theory that $\mathbf{\Sigma}_0^1$, $\mathbf{\Pi}_1^1$ and $\mathbf{\Sigma}_2^1$ satisfy the reduction property. Whether reduction holds on the $\mathbf{\Sigma}$ or the $\mathbf{\Pi}$ side for $k \geqq 3$ is one of the central problems of the theory of projective sets and has provided much of the motivation for their study. The best result until very recently was in Addison 1959a, 1959b (working out a proposal of Gödel, 1940) that if every real is constructible in the sense of Gödel, then *Reduction* $(\mathbf{\Sigma}_k^1)$ for all $k \geqq 3$.

2.3 *Determinacy.* With each subset $A \subseteq R \times R$ we associate a game as follows: players I and II choose successively natural numbers $\alpha(0)$, $\beta(0)$, $\alpha(1)$, $\beta(1), \dots$ and if $(\alpha, \beta) \in A$, I wins, if $(\alpha, \beta) \notin A$, II wins. A *strategy* for player I is a real σ (utilized as a function on finite sequences of integers to integers) which tells I how to play when II plays any real β. We let

$$\sigma * [\beta] = \alpha \text{ where } \alpha(n) = \sigma(\bar{\beta}(n)).$$

Similarly for player II we put

$$[\alpha] * \tau = \beta \text{ where } \beta(n) = \tau(\bar{\alpha}(n+1)).$$

A strategy σ is *winning* for I if for all β, $(\sigma * [\beta], \beta) \in A$; τ is winning for II if for all α, $(\alpha, [\alpha] * \tau) \notin A$. The set A is *determined* if either I or II has a winning strategy—it is trivial that they cannot both have winning strategies.

Infinite games of this type were introduced in Gale-Stewart 1953, where it was shown that every closed set is determined and that there exist non-determined sets. The relevance of determinacy for definability problems in analysis was emphasized by Mycielski and Steinhaus in the early sixties, cf. Mycielski 1964 for references.

The proof that non-determined sets exist utilizes the axiom of choice, probably in an essential way. Hence Zermelo-Fraenkel set theory (without choice) may be consistent with the following proposition.

Axiom of Determinacy, AD. Every subset of $R \times R$ *is determined.*

Fortunately AD implies (trivially) the *Countable axiom of choice for sets of reals,*

$$\forall n \exists \alpha (n, \alpha) \in A \;\Rightarrow\; \exists \alpha \forall n (n, (\alpha)_n) \in A \qquad (A \subseteq \omega \times R)$$

which is indispensable in analysis. AD also implies several "desirable falsehoods", e.g. that every set of real numbers is Lebesgue measurable and (closer to our subject) that every wellorderable subset of R is countable. Such results led Steinhaus to propose that we replace the axiom of choice by AD in our thinking about sets.

A more persuasive incentive for studying consequences of AD is the hope that we may find a natural class of sets which contains R and satisfies AD—this was first suggested by Mycielski. Recently Solovay has been conjecturing that

$$L[R] = \textit{the class of sets constructible from } R$$

satisfies AD and that this may be provable from suitable axioms of infinity. A very powerful argument for Solovay's conjecture is the recent theorem of Martin: *if there exist measurable cardinals, then every* $\mathbf{\Pi}^1_1$ *set is determined.* (However a theorem of Silver together with the results mentioned in 2.4 below shows that the existence of measurable cardinals does not imply that every $\mathbf{\Delta}^1_2$ set is determined, unless through an inconsistency.)

Addison-Moschovakis 1968 proposed that *every subset of* $R \times R$ *ordinal definable from a real is determined.* At this time we have no lead for attempting to prove this assertion from plausible axioms, so we must view the proposal only as a challenge to derive consequences from it that contradict our basic intuitions about sets.

Sometimes we can obtain interesting results by assuming that only sets in a certain class Λ are determined. Put

$$\textit{Determinacy } (\Lambda)\text{: } \textit{every } A \in \Lambda \textit{ is determined},$$

where as usual $A \in \Lambda$ for $A \subseteq R \times R$ means $A \in [R \times R]\Lambda$. Consequences of *Determinacy*(Λ) are particularly pleasing when the sets in Λ are definable so that the hypothesis is plausible.

2.4 *Prewellorderings.* A *prewellordering of a set A (with field A)* is a relation $\leqq$ on A such that for some ordinal ξ, some surjection $\phi: A \twoheadrightarrow \xi$ and all $x, y \in A$,

$$x \leqq y \Leftrightarrow \phi(x) \leqq \phi(y),$$

where on the right $\leqq$ is the ordering relation on ordinals. It is immediate that such ϕ and ξ are unique when they exist; we call ξ the *length* of $\leqq$ and ϕ the *canonical surjection* (of $\leqq$).

A relation $\leqq$ is a prewellordering if and only if it is reflexive, transitive, connected and well founded — from being a wellordering it only lacks antisymmetry. It often happens that we can prove the first three of these properties for some $\leqq$, but instead of well foundedness we only have the apparently weaker *lack of infinite descending chains,* i.e.

$$\forall n[x_{n+1} \leqq x_n] \Rightarrow \exists n[x_n \leqq x_{n+1}],$$

where $\{x_n\}_{n=0}^{\infty}$ varies over all sequences in A. In order to infer that $\leqq$ is a prewellordering, we then need the axiom of *dependent choices* — in our case, when $A \subseteq \mathfrak{X}$ for some $\mathfrak{X}$, the following version for sets of reals is enough:

Dependent Choices, DC. For each $A \subseteq R \times R$,

$$\forall\alpha\exists\beta(\alpha, \beta) \in A \Rightarrow \exists\alpha\forall n((\alpha)_n, (\alpha)_{n+1}) \in A.$$

This follows from the axiom of choice and it is not known whether it follows from AD, or whether it is consistent with AD. However any class of sets that contains R must satisfy DC, since DC holds in the universe, hence any "natural" model of AD (hopefully $L[R]$) will also satisfy DC.

Consider the following property of a class Γ parametrized by G.

Prewellordering (Γ). *There is a prewellordering $\leqq$ on G and relations $\underset{\cdot}{\leqq}$ and $\dot{\leqq}$ in Γ and $\neg\Gamma$ respectively, such that*

(2–2) $$w \in G \Rightarrow \forall z\{z \underset{\cdot}{\leqq} w \Leftrightarrow z \dot{\leqq} w \Leftrightarrow [z \in G \,\&\, z \leqq w]\}.$$

If G' is also universal for Γ, then

$$(\alpha, \beta) \in G' \Leftrightarrow (\gamma_0, f(\alpha, \beta)) \in G$$

for some recursive f and some γ_0. If we define

$$(\alpha_1, \beta_1) \leqq' (\alpha_2, \beta_2) \Leftrightarrow (\gamma_0, f(\alpha_1, \beta_1)) \leqq (\gamma_0, f(\alpha_2, \beta_2)),$$

we can easily verify that the conditions of *Prewellordering* (Γ) are satisfied with $\leqq'$ and suitable $\underset{\cdot}{\leqq}', \dot{\leqq}'$. Thus the prewellordering property depends on Γ alone and not on any particular universal set.

It is easy to verify that if Γ is parametrized and closed under recursive substitution and &, then

$$\textit{Prewellordering}\ (\Gamma) \Rightarrow \textit{Reduction}\ (\Gamma).$$

Actually *Prewellordering* (Γ) implies many interesting structure results about Γ, especially when Γ satisfies sufficiently strong closure conditions.

One may say that

(I) *Prewellordering* ($\boldsymbol{\Sigma}^1_0$), *Prewellordering* ($\boldsymbol{\Pi}^1_1$)

were known classically. (Actually *Prewellordering* ($\boldsymbol{\Sigma}^1_0$) was only noticed by Addison in 1968.) We formulated the prewellordering property in 1964 in order to prove *Prewellordering* ($\boldsymbol{\Sigma}^1_2$) and thus lift in an elegant manner the theory of $\boldsymbol{\Pi}^1_1$ to $\boldsymbol{\Sigma}^1_2$. (One of our results was the construction of a hierarchy for $[\omega]\Delta^1_2$, which we then found had been achieved by Suzuki 1964.) This construction appeared in Rogers 1967 where it was shown (in effect) that

(II) *if* Γ *is parametrized, closed under recursive substitutions,* &, $\vee$, $\forall\alpha$ *and Prewellordering* (Γ), *then Prewellordering* ($\Sigma^1\Gamma$).

The prewellordering property was the key to the development of the theory of semi-hyperanalytic sets in Moschovakis 1967 and the theory of semi-hyperprojective sets in Moschovakis 1969. In each of these cases we can lift much of the theory of $\boldsymbol{\Pi}^1_1$ to these classes — and we can do more because of the stronger closure properties that we can utilize.

The arguments of Addison, 1959a, suffice to show

(III) *if* Γ *is parametrized, closed under recursive substitution,* &, $\vee$, $\exists m$, $\forall m$ *and* $\forall\alpha$ *and some relation in* Γ *wellorders R with order type* $\aleph_1$, *then Prewellordering* ($\Sigma^1\Gamma$).

These results imply that if every real is constructible in the sense of Gödel, then for each $k \geqq 2$, *Prewellordering* ($\boldsymbol{\Sigma}^1_k$).

Determinacy enters this picture via the next theorem, the main result in Addison-Moschovakis 1968 and Martin 1968.

(IV) *If* Γ *is parametrized, closed under continuous substitution,* &, $\vee$, $\exists m$, $\forall m$ *and* $\exists\alpha$, *if Prewellordering* (Γ), *DC and Determinacy* ($\Gamma \cap \neg\Gamma$), *then Prewellordering* ($\Pi^1\Gamma$).

If we assume *DC* and the determinacy of all projective sets, then (I), (II) and (IV) imply *Prewellordering* ($\boldsymbol{\Sigma}^1_k$) for all even k and *Prewellordering* ($\boldsymbol{\Pi}^1_k$) for all odd k. Since the prewellordering property cannot hold on both the $\boldsymbol{\Sigma}$ and the $\boldsymbol{\Pi}$ side for the same k (because it implies reduction), this picture of the projective hierarchy is radically different from the picture when we assume that every real is constructible. Which is the correct picture is perhaps not absolutely clear yet, but it is fair to say that many people working in this area and prone to speak about truth in set theory (ourselves included) tend to favor the alternating picture. In fact the most persuasive argument for accepting projective determinacy (aside from Martin's proof of Determinacy ($\mathbf{\Pi}_1^1$)) is the naturalness of the known proofs of (IV), both Martin's and ours.

One of the central open problems in the theory of definability on the continuum is whether the prewellordering property holds for $\mathbf{\Sigma}_1^2$ or $\mathbf{\Pi}_1^2$. Trivial extensions of the method used for (III) show that if every set of reals is in $L[R]$, then *Prewellordering* $(\mathbf{\Sigma}_1)$, but we are interested in answering this question using axioms that do not restrict our conception of set. It is not unreasonable to suppose that *Prewellordering* $(\mathbf{\Sigma}_1^2)$ may be provable in Zermelo-Fraenkel.

§3. The basic lemmas. Let Λ be a class of subsets of R, to avoid trivialities assume that Λ contains all singletons and is closed under continuous substitution, & and $\neg$. An ordinal ξ is *realized* in Λ if there is a prewellordering $\leqq$ of some subset of R in Λ (i.e. in $[R \times R]\Lambda$) with length ξ. The assumed closure properties of Λ imply

$$\xi \textit{ is realized in } \Lambda \ \& \ \zeta < \xi \Rightarrow \zeta \textit{ is realized in } \Lambda,$$

$$\xi \textit{ is realized in } \Lambda \Rightarrow \xi + 1 \textit{ is realized in } \Lambda,$$

$$\xi > 0 \ \& \ \xi \textit{ is realized in } \Lambda \Rightarrow \xi \textit{ is the length of some prewellordering of } R \textit{ in } \Lambda.$$

It is also clear that the same ordinals are realized if we allow prewellorderings on subsets of any product space $\mathfrak{X}$. Put

$$o(\Lambda) = \textit{supremum}\ \{\xi : \xi \textit{ is realized in } \Lambda\}.$$

We wish to study $o(\Lambda)$, especially when Λ satisfies nontrivial closure conditions, e.g. $\Lambda = \mathbf{\Delta}_k^1, \mathbf{\Delta}_1^2, {}^R2$. The basic theory in which we work is Zermelo-Fraenkel without choice but with the countable axiom of choice for sets of reals. *All results stated thus far in this paper are provable in this theory.* We shall often assume AD and sometimes DC, but then will mark the theorems accordingly. To verify that $o(\Lambda)$ is always defined even without the axiom of choice, notice that

$$\begin{aligned} o({}^R2) &= \textit{supremum}\ \{\xi : \textit{there exists a surjection } \phi : R \twoheadrightarrow \xi\} \\ &= \textit{supremum}\ \{\xi : \textit{there exists an injection } \psi : \xi \rightarrow {}^R2\}; \end{aligned}$$

now the second class in braces is bounded by a classical argument of Hartog.

In this section we give the basic new constructions of this paper.

A formula of the language of analysis is Σ^1_1 if it is of the form

$$\exists\alpha_1\exists\alpha_2 \ldots \exists\alpha_n\theta\,, \tag{3–1}$$

where θ has no real quantifiers. These are the formulas from which we get the Σ^1_1 subsets of the product spaces $\mathfrak{X}$, by fixing the values of some of the real variables and considering the remaining free variables as a "vector" variable over $\mathfrak{X}$. Up to equivalence in the standard interpretations these formulas satisfy all the closure properties of Σ^1_1 sets.

Let χ be a partial function on R to ω. We extend the language of analysis by adding prime formulas of the form "$\chi(\delta) \simeq m$", which in the standard interpretation mean "$\chi(\delta)$ is defined and $= m$". Now $\Sigma^1_1(\chi)$ formulas are those of the form (3–1) where θ has no real quantifiers and *only positive occurrences of subformulas of the form* $\chi(\delta) \simeq m$. We let $\Sigma_1(\chi)$ be the collection of subsets of R which are definable by $\Sigma^1_1(\chi)$ formulas of this type, after we fix the values of all the variables but one real variable.

If χ is totally defined, then the restriction to positive occurrences of subformulas of the form $\chi(\delta) \simeq m$ is not essential, since

$$\neg(\chi(\delta) \simeq m) \Leftrightarrow \exists n[n \neq m \,\&\, \chi(\delta) \simeq n].$$

LEMMA 1. *For each $\Sigma^1_1(\chi)$ formula Θ there is a Σ^1_1 formula $\Psi(\alpha,\beta)$ (with no occurrences of χ) such that*

$$\Theta \Leftrightarrow \exists\alpha\exists\beta\{\forall n[\chi(\alpha)_n \simeq \beta(n)] \,\&\, \Psi(\alpha,\beta)\}, \tag{3–2}$$

i.e. the universal closure of (3–2) *is true in the standard interpretation.*

Proof. It is enough to prove the lemma for Θ that have no real quantifiers, since the result then follows by simple quantification. These are built up from formulas with no occurrences of χ and formulas of the form $\chi(\delta) \simeq m$ by the positive operations &, $\vee$, $\exists m$, $\forall m$, and the proof is by induction over this construction. One of the basis cases is handled by

$$\chi(\delta) \simeq m \Leftrightarrow \exists\alpha\exists\beta\{\forall n[\chi((\alpha)_n) \simeq \beta(n)] \,\&\, (\alpha)_0 = \delta \,\&\, \beta(0) = m\}.$$

To treat the most complicated case in the inductive step, suppose $\Theta \Leftrightarrow \forall t\Theta^*(t)$, where by induction hypothesis,

$$\Theta^*(t) \Leftrightarrow \exists\alpha\exists\beta\{\forall n[\chi((\alpha)_n) \simeq \beta(n)] \,\&\, \Psi^*(\alpha,\beta,t)\}.$$

Now the countable axiom of choice for sets of reals implies that

$$\begin{aligned}\forall t\Theta^*(t) \Leftrightarrow \exists\gamma\exists\delta\{&\forall n[\chi((\gamma)_n) \simeq \delta(n)]\\ &\&\ \forall t\exists\alpha\exists\beta\{\forall n[(\alpha)_n = (\gamma)_{\langle t,n\rangle}]\\ &\&\ \forall n[\beta(n) = \delta(\langle t,n\rangle)]\\ &\&\ \Psi^*(\alpha,\beta,t)\}\},\end{aligned}$$

from which the desired conclusion follows by using the closure properties (up to equivalence) of Σ_1^1 formulas.

From this lemma we obtain immediately

LEMMA 2. *Let $G \subseteq R^4$ be universal for $[R^3]\Sigma_1^1$, the ternary relations in $\boldsymbol{\Sigma}_1^1$, for each partial $\chi: R \to \omega$ put*

$$(3\text{–}3) \qquad (\varepsilon, \delta) \in G(\chi) \Leftrightarrow \exists\alpha\exists\beta\{\forall n[\chi((\alpha)_n) \simeq \beta(n)] \;\&\; (\varepsilon, \delta, \alpha, \beta) \in G\};$$

then $G(\chi)$ is universal for $\boldsymbol{\Sigma}_1^1(\chi)$.

Let $\leqq$ be a relation with field a subset of some $\mathfrak{X}$ with *type* $\mathfrak{X} = 1$, let $\pi: R \to \mathfrak{X}$ be the canonical recursive homeomorphism. The canonical partial $\chi: R \to \omega$ associated with $\leqq$ is the characteristic function of $\leqq$, restricted to the domain of $\leqq$ and carried over to R for convenience,

$$(3\text{–}4) \qquad \begin{aligned} \chi(\alpha) \simeq n \Leftrightarrow {} & [\pi(\alpha)_0 \leqq \pi(\alpha)_1 \;\&\; n = 0] \\ \vee {} & [\pi(\alpha)_0 \leqq \pi(\alpha)_0 \;\&\; \pi(\alpha)_1 \leqq \pi(\alpha)_1 \\ & \qquad \& \neg[\pi(\alpha)_0 \leqq \pi(\alpha)_1] \\ & \qquad \& \; n = 1]. \end{aligned}$$

We shall be concerned with the class

$$(3.5) \qquad \boldsymbol{\Sigma}_1^1(\leqq) = \boldsymbol{\Sigma}_1^1(\chi),$$

for this canonically defined χ, especially when $\leqq$ is a prewellordering.

If the length of $\leqq$ is ξ and ϕ: *Field*$(\leqq) \twoheadrightarrow \xi$ is the canonical surjection, then ϕ gives a "coding" of ξ in the space $\mathfrak{X}$, i.e. we can think of each $x \in Field(\leqq)$ such that $\phi(x) = \eta < \xi$ as a code or name for η. If $f:\xi \to {}^{\mathfrak{Y}}2$ is a function on ξ to subsets of some $\mathfrak{Y}$, we can represent it by a subset of $\mathfrak{X} \times \mathfrak{Y}$ as follows:

$$Cod(f; \leqq) = \{(x, y): x \leqq x \;\&\; y \in f(\phi(x))\}.$$

Suppose $f: \xi \to {}^{\mathfrak{Y}}2$ is a function. A *choice subfunction* of f is any $g: \xi \to {}^{\mathfrak{Y}}2$ such that for all $\eta < \xi$,

$$\begin{aligned} g(\eta) &\subseteq f(\eta), \\ f(\eta) \neq \emptyset &\Rightarrow g(\eta) \neq \emptyset. \end{aligned}$$

The interesting case is when for each $\eta < \xi, f(\eta) \neq \emptyset$, when $g(\eta)$ "chooses" a nonempty subset of each $f(\eta)$.

LEMMA 3 (Main Lemma). *Assume AD. Let $\leqq$ be a prewellordering with field a subset of some $\mathfrak{X}$ and length ξ, let $f: \xi \to {}^{\mathfrak{Y}}2$ be a function.*

Then there exists a choice subfunction g of f such that $Cod(g; \leqq)$ is a $\Sigma_1^1(\leqq)$ subset of $\mathfrak{X} \times \mathfrak{Y}$.

Proof. For each $\zeta \leqq \xi$, let f_ζ be the restriction of f to ζ, modified to give $\emptyset$ on $\xi - \zeta$,

$$\begin{aligned} f_\zeta(\eta) &= f(\eta) \quad \text{if } \eta < \zeta \\ &= \emptyset \quad \text{if } \zeta \leqq \eta < \xi . \end{aligned}$$

Suppose there is some $\zeta \leqq \xi$ such that f_ζ does not have a choice subfunction with *Cod* in $\Sigma_1^1(\leqq)$, let λ be the smallest such ζ. The lemma will be proved if we can deduce a contradiction from this assumption.

Let $\phi: Field(\leqq) \twoheadrightarrow \xi$ be the canonical surjection.

First we argue that λ is a limit ordinal; because if $\lambda = \zeta + 1$ and g_ζ is a choice subfunction of f_ζ, then either $f(\zeta) = \emptyset$ and g_ζ is a choice subfunction of f_λ or there is some $y_0 \in f(\zeta)$ and

$$g_\lambda = g_\zeta - \{(\zeta, \emptyset)\} \cup \{(\zeta, \{y_0\})\}$$

is a choice subfunction of f_λ with

$$(x, y) \in Cod(g_\lambda; \leqq) \Leftrightarrow (x, y) \in Cod(g_\zeta; \leqq) \vee [\phi(x) = \zeta \;\&\; y = y_0].$$

If $x_0 \in Field(\leqq)$ is chosen so that $\phi(x_0) = \zeta$, then

$$\phi(x) = \zeta \Leftrightarrow x \leqq x_0 \;\&\; x_0 \leqq x,$$

so that $Cod(g_\lambda; \leqq)$ is in $\Sigma_1^1(\leqq)$ contradicting the choice of λ.

By Lemma 2, $\Sigma_1^1(\leqq)$ is parametrized, hence the class $[\mathfrak{X} \times \mathfrak{Y}]\Sigma_1^1(\leqq)$ of $\Sigma_1^1(\leqq)$ subsets of $\mathfrak{X} \times \mathfrak{Y}$ is parametrized, let $G \subseteq R \times \mathfrak{X} \times \mathfrak{Y}$ be a fixed universal set for it. As usually,

$$G_\alpha = \{(x, y) : (\alpha, x, y) \in G\}.$$

Consider the following two person game. If I plays α and II plays β, then

$$\begin{aligned} \text{II } \textit{wins} \Leftrightarrow \neg\, &(\exists \eta)[g_\eta \textit{ is a choice subfunction of } f_\eta \;\&\; G_\alpha = Cod(g_\eta; \leqq] \\ &\vee (\exists \eta < \lambda)(\exists \zeta < \lambda)(\exists g_\eta)(\exists g_\zeta)[g_\eta \textit{ is a choice subfunction of } f_\eta \\ &\& \; g_\zeta \textit{ is a choice subfunction of } f_\zeta \;\&\; \eta < \zeta \\ &\& \; G_\alpha = Cod(g_\eta; \leqq) \;\&\; G_\beta = Cod(g_\zeta; \leqq)]. \end{aligned}$$

If we think of α as a *code* of a function g when $G_\alpha = Cod(g; \leqq)$, then II wins if either I does not code a choice subfunction of an initial segment of f or I does, and II codes a choice subfunction of a longer initial segment of f.

By AD, the game is determined.

Case 1. I *has a winning strategy* σ. Now for each β there is some $\eta = \eta(\beta)$ and some $g_{\eta(\beta)}$, a choice subfunction of $f_{\eta(\beta)}$, so that

$$G_{\sigma*[\beta]} = Cod(g_{\eta(\beta)}; \leqq).$$

If *supremum* $\{\eta(\beta): \beta \in R\} < \zeta < \lambda$ for some ζ, then II can win against this σ by playing β so that $G_\beta = Cod(g_\zeta; \leqq)$, for some choice subfunction g_ζ of f_ζ — such a β exists by the choice of λ. Since λ is limit, the other alternative is *supremum* $\{\eta(\beta): \beta \in R\} = \lambda$. Put

$$g_\lambda(\zeta) = \bigcup_\beta g_{\eta(\beta)}(\zeta);$$

now g_λ is clearly a choice subfunction of f_λ and

$$(x, y) \in Cod(g_\lambda; \leqq) \Leftrightarrow \exists\beta[(\sigma*[\beta], x, y) \in G],$$

which implies that $Cod(g_\lambda; x)$ is in $\Sigma_1^1(\leqq)$ contradicting the choice of λ.

Case 2. II *has a winning strategy* τ. For each $\varepsilon \in R$, $w \in \mathfrak{X}$, consider the set $A_{\varepsilon,w} \subseteq \mathfrak{X} \times \mathfrak{Y}$ defined by

$$(x, y) \in A_{\varepsilon,w} \Leftrightarrow w \leqq w \;\&\; \exists z[z \leqq w \;\&\; \neg(w \leqq z) \;\&\; \{\varepsilon\}(z) \textit{ is defined}$$
$$\& \; (\{\varepsilon\}(z), x, y) \in G],$$

where $\{\varepsilon\}(z) \simeq \{\varepsilon\}^{\mathfrak{X},R}(z)$, i.e. we think of ε as a code for a continuous partial function on $\mathfrak{X}$ to R. It is clear from the closure properties of $\Sigma_1^1(\leqq)$ that each $A_{\varepsilon,w}$ is in $\Sigma_1^1(\leqq)$ — notice that $\neg(w \leqq z) \Leftrightarrow \exists\alpha[\pi(\alpha)_0 = w \;\&\; \pi(\alpha)_1 = z \;\&\; \chi(\alpha) \simeq 1]$ when χ is the canonical partial function on R to ω associated with $\leqq$, and $w \leqq w$, $z \leqq w$, i.e. $z \in Field(\leqq)$. In fact, there is a recursive function $g: R \times \mathfrak{X} \to R$ such that for each ε, w,

$$A_{\varepsilon,w} = G_{g(\varepsilon,w)} = \{(x, y): (g(\varepsilon, w), x, y) \in G\}.$$

This follows by recalling that G was defined explicitly by a $\Sigma_1^1(\chi)$ formula and that the closure of $\Sigma_1^1(\chi)$ under the operations &, $\vee$, $\exists\alpha$, etc. follows from simple explicit manipulations of $\Sigma_1^1(\chi)$ formulas which can be easily made to correspond to recursive operations on the codes.

The recursion theorem for continuous partial functions implies that there is a fixed real ε^* such that for all $w \in \mathfrak{X}$,

$$\{\varepsilon^*\}(w) \simeq [g(\varepsilon^*, w)] * \tau,$$

where the strong equality $\simeq$ is the same as $=$ here, since g is totally defined on $R \times \mathfrak{X}$, so that $[g(\varepsilon^*, w)] * \tau$ is always a real.

SUBLEMMA. *For each* $w \in Field(\leqq)$ *there is some* $\zeta = \zeta(w)$ *and some choice subfunction* $g_{\zeta(w)}$ *of* $f_{\zeta(w)}$ *such that*

$$\zeta(w) > \phi(w)$$

and

$$G_{\{\varepsilon^*\}(w)} = Cod(g_{\zeta(w)} \leqq).$$

Proof of the sublemma is by transfinite induction on $\phi(w)$. If both of the assertions hold for all $z \in Field(\leqq)$ with $\phi(z) < \phi(w)$, then easily

$$\begin{aligned} A_{\varepsilon^*,w} &= \bigcup \{Cod(g_{\zeta(z)}; \leqq): \phi(z) < \phi(w)\} \\ &= Cod(g_\eta; \leqq), \end{aligned}$$

where

$$\eta = supremum\{\zeta(z): \phi(z) < \phi(w)\}$$

and for $\zeta < \eta$,

$$g_\eta(\zeta) = \bigcup \{g_{\zeta(z)}(\zeta): \phi(z) < \phi(w)\}.$$

Clearly g_η is a choice subfunction of f_η; hence if $\eta = \lambda$, we have already obtained a contradiction, since $A_{\varepsilon^*,w}$ is $\Sigma^1_1(\leqq)$. If $\eta < \lambda$, we have in any case that $\eta \geqq \phi(w)$, since by induction hypothesis for each z with $\phi(z) < \phi(w)$, $\zeta(z) > \phi(z)$. Now

$$A_{\varepsilon^*,w} = G_{g(\varepsilon^*,w)} = Cod(g_\eta; \leqq)$$

and the choice of τ implies that

$$G_{[g(\varepsilon^*,w)]*\tau} = Cod(g_\zeta; \leqq)$$

for some $\zeta > \eta \geqq \phi(w)$ and some choice subfunction g_ζ of f_ζ. Since $[g(\varepsilon^*, w)] * \tau = \{\varepsilon^*\}(w)$ by the choice of ε^*, the proof of the sublemma is complete.

If we now put

$$g_\lambda(\zeta) = \bigcup_{w \in Field(\leqq)} g_{\zeta(w)}(\zeta),$$

then easily g_λ is a choice subfunction of f_λ and

$$(x, y) \in Cod(g_\lambda; \leqq) \Leftrightarrow \exists w[w \leqq w \;\&\; (\{\varepsilon^*\}(w), x, y) \in G],$$

so that $Cod(g_\lambda; \leqq)$ is in $\Sigma^1_1(\leqq)$ contradicting the choice of λ and completing the proof of the lemma.

Let $\leqq$ be a prewellordering with length ξ and canonical surjection $\phi: Field(\leqq) \twoheadrightarrow \xi$. Let $f: \xi^n \to {}^{\mathfrak{Y}}2$ be an n-ary function on ξ to subsets of $\mathfrak{Y}$. By analogy with unary functions put

$$Cod(f; \leqq) = \{(x_1, \ldots, x_n, y): x_1, \ldots, x_n \in Field(\leqq) \;\&\; y \in f(\phi(x_1), \ldots, \phi(x_n))\}.$$

Similarly for any n-ary relation P on ξ and any subset A of ξ, put

$$\begin{aligned} Cod(P; \leqq) &= \{(x_1, \ldots, x_n): x_1, \ldots, x_n \in Field(\leqq) \;\&\; P(\phi(x_1), \ldots, \phi(x_n))\}, \\ Cod(A; \leqq) &= \{x : x \in Field(\leqq) \;\&\; \phi(x) \in A\}. \end{aligned}$$

If $f\colon \xi^n \to {}^{\mathfrak{Y}}2$, then a *choice subfunction* of f is any $g\colon \xi^n \to {}^{\mathfrak{Y}}2$ such that for all $\eta_1, \cdots, \eta_n < \xi$,

$$g(\eta_1, \ldots, \eta_n) \subseteq f(\eta_1, \ldots, \eta_n);$$

$$f(\eta_1, \ldots, \eta_n) \neq \emptyset \to g(\eta_1, \ldots, \eta_n) \neq \emptyset.$$

LEMMA 4. *Assume AD. Let $\leqq$ be a prewellordering with field a subset of some $\mathfrak{X}$ and length ξ, let $f\colon \xi^n \to {}^{\mathfrak{Y}}2$ be a function. Then there exists a choice subfunction g of f such that $Cod(g; \leqq)$ is a $\Sigma^1_1(\leqq)$ subset of $\mathfrak{X} \times \mathfrak{Y}$.*

Proof. We know the lemma for unary functions, so to proceed inductively let us assume it for n-ary functions and let f be an $n+1$-ary function on ξ to ${}^{\mathfrak{Y}}2$. For each $\eta < \xi$, put

$$f^\eta(\eta_1, \ldots, \eta_n) = f(\eta_1, \ldots, \eta_n, \eta)$$

and set

$$f^*(\eta) = \{\alpha\colon \textit{for some choice subfunction } g^\eta \text{ of } f^\eta, G_\alpha = Cod(g^\eta, \leqq)\},$$

where G is universal for the $\Sigma^1_1(\leqq)$ subsets of $(\mathfrak{X})^n \times \mathfrak{Y}$. By Lemma 3, f^* has a choice subfunction g^* with $Cod(g^*; \leqq)$ in $\Sigma^1_1(\leqq)$. Moreover, if

$$(x_1, \ldots, x_n, x, y) \in A \Leftrightarrow \exists\alpha[(x, \alpha) \in Cod(g^*; \leqq) \mathbin{\&} (\alpha, x_1, \ldots, x_n, y) \in G],$$

then A is $\Sigma^1_1(\leqq)$ and it is easy to verify that

$$A = Cod(g; \leqq)$$

for some choice subfunction g of f.

LEMMA 5. *Assume AD. Let $\leqq$ be a prewellordering with field a subset of some $\mathfrak{X}$ and length ξ, let P be an n-ary relation on ξ, A a subset of ξ. Then $Cod(P; \leqq)$, $Cod(A; \leqq)$ are $\Sigma^1_1(\leqq)$.*

Proof. The part about relations follows easily from Lemma 4 if we associate with n-ary P on ξ the function

$$\begin{aligned} f(\eta_1, \ldots, \eta_n) &= \{\alpha_0\} \text{ if } P(\eta_1, \ldots, \eta_n), \\ &= \{\alpha_1\} \text{ if } \neg P(\eta_1, \ldots, \eta_n), \end{aligned}$$

where α_0, α_1 are two fixed distinct reals. Now the only choice subfunction of f is f itself, so by Lemma 4 $Cod(f; \leqq)$ is $\Sigma^1_1(\leqq)$ and

$$(x_1, \ldots, x_n) \in Cod(P; \leqq) \Leftrightarrow (x_1, \ldots, x_n, \alpha_0) \in Cod(f; \leqq).$$

The part about sets follows trivially from that about relations.

One interesting application of Lemma 3 is given in the next lemma.

LEMMA 6. *Assume AD. Let $\leqq$ be a prewellordering with field a subset of some $\mathfrak{X}$ and length ξ, let χ be the partial function associated with $\leqq$ via (3–4). Let Γ be a class of sets containing all singletons, parametrized and closed under continuous substitution (preimages), &, $\vee$, $\exists m$, $\forall m$ and $\exists\alpha$ and containing $\{(\delta, m): \chi(\delta) \simeq m\}$. Suppose*

$$A = \bigcup_{\eta<\xi} A_\eta,$$

where each A_η is a Γ subset of some $\mathfrak{Y}$. Then $A \in \Gamma$.

Proof. Let $G \subseteq \mathcal{N} \times \mathfrak{Y}$ be universal for the class $[\mathfrak{Y}]\Gamma$ of Γ subsets of $\mathfrak{Y}$. Put

$$f(\eta) = \{\alpha: G_\alpha = A_\eta\}$$

and by Lemma 3 choose $g: \xi \to {}^R 2$, a choice subfunction of f such that $Cod(g; \leqq) \in \Sigma_1^1(\leqq)$. Now the closure properties of Γ imply that $Cod(g; \leqq) \in \Gamma$ and we have

$$y \in A \Leftrightarrow \exists x \exists \alpha[(x, \alpha) \in Cod(g; \leqq) \;\&\; (\alpha, y) \in G],$$

so that $A \in \Gamma$.

§4. How large is $o({}^R 2)$. In set theory with the axiom of choice $2^{\aleph_0}$ is an aleph and

$$o({}^R 2) = (2^{\aleph_0})^+ = \textit{first aleph greater than } 2^{\aleph_0}.$$

AD implies that R cannot be wellordered so that $2^{\aleph_0}$ is not an aleph and the formula above makes no sense. However $o({}^R 2)$ surely is an aleph and it seems to give a reasonable measure of the size of R relative to the size of ordinals.

It turns out that in the context of *AD*, $o({}^R 2)$ is utterly huge. In this section we give a few results that suggest this. The first answers a question of Solovay.†

† Results of this type were first obtained (to the best of our knowledge) by H. Friedman and ourselves, independently, in the winter and spring of 1968. The main result at that time was that (with *AD*) many alephs which can be approached from below (e.g. $\aleph_{\aleph_1}$) are order types of prewellorderings of R. Friedman's results gave additional information about the subsets of these alephs, e.g. that each subset of $\aleph_{\aleph_1}$ is definable from a real. Our chief result was that (with *AD* and *DC*) each $\aleph_k$ is the order type of a projective prewellordering of R and that larger ordinals, like $\aleph_1$, are order types of hyperanalytic prewellorderings.

In our original proofs we only used *Solovay games*, like that in the proof of Theorem 6, i.e. games whose definition guarantees that player I cannot win; such games were first utilized by Solovay in his proof that (with *AD*) every subset of $\aleph_1$ is "Π_1^1 in the codes". Friedman's early contribution was a technique of utilizing games where it is not clear which player wins, by proving the desired result by cases; we call games used in such proofs *Friedman games*. The proof of our key Lemma 3 uses Friedman games and was obtained after we learned this technique from Friedman; we like to think of it as the product of the marriage of Friedman's technique with our method of introducing "coded classes of sets" and applying the recursion theorem.

Friedman's results will appear in a joint Friedman-Solovay paper with tentative title "*Large ordinals and the axiom of determinateness*".

THEOREM 1. *Assume AD. If there is a surjection $\phi: R \twoheadrightarrow \xi$, then there is a surjection $\phi^*: R \twoheadrightarrow {}^{\xi}2$.*
(The hypothesis is trivially equivalent to $\xi < o({}^R2)$ for $\xi \neq 0$.)

Proof. Assume $\phi: R \twoheadrightarrow \xi$, let $\leqq$ be the prewellordering with length ξ defined on R by ϕ

$$\alpha \leqq \beta \Leftrightarrow \phi(\alpha) \leqq \phi(\beta).$$

If $A \subseteq \xi$ then by Lemma 5 $Cod(A; \leqq)$ is $\boldsymbol{\Sigma}^1_1(\leqq)$; thus if G is a universal set for $\boldsymbol{\Sigma}^1_1(\leqq)$ we can define a surjection ϕ^* by

$$\begin{aligned} \phi^*(\alpha) &= A \quad \textit{if } G_\alpha = Cod(A; \leqq) \textit{ for some } A \subseteq \xi, \\ &= \emptyset \quad \textit{otherwise.} \end{aligned}$$

COROLLARY 1.1. (Friedman) *Assume AD. If $\xi < o({}^R2)$ then $\xi^+ < o({}^R2)$.*

Proof. Using Lemma 5 again, but the part about binary relations, we show as in the theorem that if $\xi < o({}^R2)$ then there is a surjection $\psi: R \twoheadrightarrow {}^{(\xi \times \xi)}2$. But there is an obvious surjection $\chi: {}^{(\xi \times \xi)}2 \twoheadrightarrow \xi^+$ given by

$$\begin{aligned} \chi(A) &= \textit{order type of } A, \textit{ if } A \subseteq \xi \times \xi \textit{ is a wellordering,} \\ &= 0 \quad \textit{otherwise,} \end{aligned}$$

so there is a surjection $\chi \circ \psi: R \twoheadrightarrow \xi^+$, hence $\xi^+ < o({}^R2)$.

COROLLARY 1.2. *Assume AD. If there is a surjection $\phi: R \twoheadrightarrow \xi$ which is ordinal definable from a real (in $L[R]$) then there is a surjection $\psi: R \twoheadrightarrow {}^{\xi}2$ which is also ordinal definable from a real (in $L[R]$) so that each subset of ξ is ordinal definable from a real (in $L[R]$).* (The part about $L[R]$ is due to Solovay.)

Proof is immediate from the proof of the theorem since the ψ we defined is ordinal definable from a real (in $L[R]$) when ϕ is.

The best lower bounds for $o({}^R2)$ that we knew before the Jerusalem meeting were cardinals λ which are π_λ in the sense of Mahlo. At the meeting Solovay told us of his results that the first Mahlo, the first fixed points of Mahlos, etc. are less than $o({}^R2)$. We prove here a theorem which gives these results (and apparently more) and which will also yield definability estimates in §7.†

† Our original proofs that *AD* implies the existence of large cardinals (inaccessibles of high order but not Mahlo) used recursion theoretic techniques and in particular the theory of hyperanalytic and hyperprojective subsets of R, in Moschovakis 1967, 1969. The proof of Theorem 2 below uses a method that we learned from Solovay in Jerusalem: to show that a cardinal with a certain property exists and is less than $o({}^R2)$, assume that no $\xi < \varkappa$ has that property and show that we can then code all ordinals less than $\varkappa$ using reals. Again, we like to think of our Theorem 2 as a child of coupling this technique with our coding methods embodied in Lemma 3. Solovay's applications of this technique will appear in the Friedman-Solovay paper mentioned in the preceding footnote.

Let L be a *denumerable* first order language, perhaps with many sorts of variables, among which we distinguish one, the *ordinal variables.* Suppose that to each ordinal λ we have assigned an L-structure $\mathfrak{A}_\lambda$, in which the ordinal variables range over the ordinals less than λ. If $A \subseteq \lambda$, we let $(\mathfrak{A}_\lambda, A)$ be the structure obtained from $\mathfrak{A}_\lambda$ by adding one set (unary predicate) of ordinals — let L' be the language suitable for these structures. If $\lambda < \kappa$, $A \subseteq \lambda$ and $B \subseteq \kappa$, put

$$(\lambda, A) \prec\prec (\kappa, B)$$

if for every formula $\theta(x_1, \ldots, x_n)$ of L' whose only free variables are ordinal variables and every n-tuple $\xi_1, \ldots, \xi_n$ or ordinals less than λ,

$$(\mathfrak{A}_\lambda, A) \models \theta(\xi_1, \ldots, \xi_n) \Leftrightarrow (\mathfrak{A}_\kappa, B) \models \theta(\xi_1, \ldots, \xi_n).$$

(From this follows that $A = \lambda \cap B$, but it need not be the case that $\mathfrak{A}_\lambda$ is a substructure of $\mathfrak{A}_\kappa$.)

We say that κ *reflects with respect to L and* $\{\mathfrak{A}_\lambda\}$ if for each $A \subseteq \kappa$, $\{\lambda : (\lambda, A \cap \lambda) \prec\prec (\kappa, A)\}$ is non-empty and unbounded in κ.

THEOREM 2. *Assume AD. Let a denumerable L and a class* $\{\mathfrak{A}_\lambda\}$ *of structures be given. There exists a regular cardinal* $\kappa < o({}^R2)$ *which reflects with respect to L and* $\{\mathfrak{A}_\lambda\}$.

We first prove a lemma.

LEMMA 7. *Let* χ_1, χ_2 *be partial functions on R to* ω, *assume* $\chi_1 \subseteq \chi_2$, *i.e.* χ_1 *is a subfunction of* χ_2. *If* $G(\chi)$ *is the canonical universal set for* $\Sigma_1^1(\chi)$ *subsets of* $\mathfrak{X}$ *defined by* (3–3) *and* (2–1). *then* $G(\chi_1) \subseteq G(\chi_2)$.

Proof. We defined $G(\chi)$ first for $\Sigma_1^1(\chi)$ subsets of R by a fixed $\Sigma_1^1(\chi)$ formula in (3–3) and then we passed to other $\mathfrak{X}$'s by (2–1), via recursive preimages. The lemma follows trivially from our requirement that $\Sigma_1^1(\chi)$ formulas have only positive occurrences of subformulas of the form $\chi(\delta) \simeq m$, so that when they are true for some χ, they are true for all extensions.

Proof of the theorem. Let us extend to arbitrary finite sequences of ordinals the canonical wellordering on pairs of ordinals,

$$\begin{aligned}(\xi_1, \ldots, \xi_n) < (\eta_1, \ldots, \eta_m) \Leftrightarrow\ & max(\xi_1, \ldots, \xi_n) < max(\eta_1, \ldots, \eta_m) \\ & \vee\ [max(\xi_1, \ldots, \xi_n) = max(\eta_1, \ldots, \eta_m)\ \&\ n < m] \\ & \vee\ max(\xi_1, \ldots, \xi_n) = max(\eta_1, \ldots, \eta_m)\ \&\ n = m \\ & \&\ (\exists i \leqq n)(\forall j < i)\ [\xi_j = \eta_j\ \&\ \xi_i < \eta_i]],\end{aligned}$$

i.e. we order sequences first by the maximum, within the same maximum by the length and within the same maximum and length lexicographically. Now the predecessors of each sequence form a set, so there exists a unique order preserving map F on the class of all ordinals to the class of all finite sequences of ordinals. It is easy to verify that when ξ is regular, $\xi > \omega$, then F restricted to ξ gives a bijection of ξ with all finite sequences of ordinals $< \xi$; we alter F slightly so that it has this property also at ω.

Let L, $\{\mathfrak{A}_\lambda\}$ be given, assume κ is an ordinal such that no regular $\xi < \kappa$ reflects with respect to L, $\{\mathfrak{A}_\lambda\}$; the theorem will follow if we can prove $\kappa < o({}^R2)$.

We define by transfinite induction a mapping $\phi : \kappa \to {}^R2$, where for each $\xi < \kappa$ we think of the elements of $\phi(\xi)$ as the *codes* for ξ. If ϕ has been defined for all $\eta < \xi$, where $\xi \leqq \kappa$, put

$$\alpha \leqq_\xi \beta \Leftrightarrow (\exists \eta < \xi)(\exists \zeta < \xi)[\alpha \in \phi(\eta) \ \& \ \beta \in \phi(\zeta) \ \& \ \eta \leqq \zeta].$$

It will turn out that $\leqq_\kappa$ is a prewellordering of length κ.

For each regular $\xi < \kappa$, each $A \subseteq \xi$, put

$Th(\xi, A) = \{\eta < \xi : F(\eta) = (m, \xi_1, \ldots, \xi_n)$ *and* m *is the Gödel number of some formula*

$$\theta(x_1, \ldots, x_n) \text{ of } L' \text{ and } (\mathfrak{A}_\xi, A) \models \theta(\xi_1, \ldots, \xi_n)\}.$$

It is immediate from the remarks about F above, that if $\eta < \xi$, η also regular and $B \subset \eta$,

$$(\eta, B) \prec\!\!\prec (\xi, A) \Leftrightarrow Th(\eta, B) \subseteq Th(\xi, A).$$

Let $\pi : \omega \times R \times R \to R$ be the canonical recursive homeomorphism. Each $\xi < \kappa$ must fall under one of the cases below.

Case 1. $\xi = 0$. *Put* $\phi(\xi) = \{\pi(1, \alpha, \alpha) : \alpha \in R\}$.

Case 2. $\xi = \zeta + 1$. *Put* $\phi(\xi) = \{\pi(2, \alpha, \alpha) : \alpha \in \phi(\zeta)\}$.

Case 3. ξ *is a singular limit ordinal. Put* $\phi(\xi) = \{\pi(3, \alpha, \beta)$: *for some* $\zeta < \xi$ *and some cofinal map* $f : \zeta \to \xi$, $\alpha \in \phi(\zeta)$ *and* $G_\beta(\leqq_\xi) = Cod(g^*; \leqq_\zeta)$, *where* g^* *is a choice subfunction of the function* $f^*(\eta) = \phi(f(\eta))\}$.

Here $G(\leqq_\zeta)$ is the canonical universal set for the class of $\mathbf{\Sigma}^1_1(\leqq_\zeta)$ subsets of $R \times R$, where $\mathbf{\Sigma}^1_1(\leqq)$ is associated with each relation $\leqq$ via (3–4) and (3–5). As usually $G_\beta(\leqq_\zeta) = \{(\gamma, \delta) : (\beta, \gamma, \delta) \in G(\leqq_\zeta)\}$.

Case 4. ξ *is a regular limit ordinal. Put* $\phi(\xi) = \{\pi(4, \alpha, \beta)$: *for some* $A \subseteq \xi$ *and some* $\zeta < \xi$ *there is no* $\lambda \geqq \zeta$ *so that* $(\lambda, A \cap \lambda) \prec\!\!\prec (\xi, A)$ *and* $\alpha \in \phi(\zeta)$ *and* $H_\beta(\leqq_\xi) = Cod(Th(\xi, A); \leqq_\xi)\}$.

Here $H(\leqq_\xi)$ is the canonical universal set for $\Sigma_1(\leqq_\xi)$ subsets of R.

We now prove by transfinite induction on $\xi < \kappa$ that:

(i) If $\pi(n, \alpha, \beta) \in \phi(\xi)$ then ξ falls under *Case* n,

(ii) $\phi(\xi) \neq \emptyset$,

(iii) $\eta < \xi \Rightarrow \phi(\eta) \cap \phi(\xi) = \emptyset$.

The induction hypothesis implies immediately that $\leqq_\zeta$ is a prewell-ordering for each $\zeta < \xi$, so we can apply the lemmas of §3. Now (i) is trivial, (ii) is trivial when ξ falls under Cases 1, 2 and follows immediately from Lemmas 3, 5 when ξ falls under Cases 3, 4 and (iii) is trivial when ξ falls under Cases 1, 2.

Proof of (iii) *when* ξ *falls under Case* 3. Suppose $\pi(3,\alpha,\beta) \in \phi(\eta) \cap \phi(\xi)$, for some $\eta < \xi$, we must derive a contradiction. By (i) of the ind. hyp., η also falls under Case 3, and using (iii) of the ind. hyp. on α we conclude the following: there is some $\zeta < \eta$ such that $\alpha \in \phi(\zeta)$ and cofinal maps $f_1 : \zeta \to \eta$, $f_2 : \zeta \to \xi$ such that $G_\beta(\leqq_\zeta) = Cod(g_1^*; \leqq_\zeta) = Cod(g_2^*; \leqq_\zeta)$, where g_1^*, g_2^* respectively are choice subfunctions of the functions $f_1^*(v) = \phi(f_1(v))$, $f_2^*(v) = \phi(f_2(v))$. Hence $g_1^* = g_2^*$, since these functions are completely determined by $Cod(g_1^*; \leqq_\zeta)$ $Cod(g_2; \leqq_\zeta)$, hence for each $v < \zeta$, $f_1^*(v) \cap f_2(v) \neq \emptyset$, i.e. $\phi(f_1(v)) \cap \phi(f_2(v)) \neq \emptyset$ so by (iii) of the ind. hyp. $f_1(v) = f_2(v)$. This implies $\eta = \xi$, contradicting $\eta < \xi$.

Proof of (iii) *when* ξ *falls under Case* 4. Suppose $\pi(4, \alpha, \beta) \in \phi(\eta) \cap \phi(\xi)$ with $\eta < \xi$, we must derive a contradiction. By (i) of the ind. hyp., η is regular, there is some $\zeta < \eta$ with $\alpha \in \phi(\zeta)$ and there are sets $A \subseteq \eta$, $B \subseteq \xi$ so that

$$H_\beta(\leqq \eta) = Cod(Th(A, \eta); \leqq_\eta),$$

$$H_\beta(\leqq_\xi) = Cod(Th(B, \xi); \leqq_\xi).$$

Let χ_η, χ_ξ be the partial functions associated with the prewellorderings $\leqq_\eta, \leqq_\xi$ via (3–4). Since $\leqq_\eta$ is an initial segemnt of $\leqq_\xi$, we have $\chi_\eta \subseteq \chi_\xi$ so that by Lemma 7,

$$Cod(Th(\eta, A); \leqq_\eta) \subseteq Cod(Th(\xi, B); \leqq_\xi).$$

But since $\leqq_\eta$ is an initial segment of $\leqq_\xi$, this implies immediately that

$$Th(\eta, A) \subseteq Th(\xi, B),$$

so that $(\eta, A) \prec\!\prec (\xi, B)$ contradicting the definition of $\phi(\xi)$.

The *dyadic second order language for one ordering* has individual variables, relation symbols $\leqq$, $=$ and $\in$, variables over subsets of the domain of individuals (unary relations) and variables over sets of pairs of individuals (binary relations). For each ordinal λ, $\mathfrak{A}_\lambda = (\lambda, {}^\lambda 2, {}^{(\lambda \times \lambda)}2, \leqq, \in)$ is the standard model of this language.

We define the Mahlo numbers in the usual way.

$$\mathfrak{M}_0 = \{\xi\colon \xi \textit{ is regular}\},$$

$$\mathfrak{M}_{\lambda+1} = \{\xi\colon \textit{each closed, unbounded } A \subseteq \xi \textit{ intersects } \mathfrak{M}_\lambda\},$$

$$\mathfrak{M}_\lambda = \bigcap\nolimits_{\eta<\lambda}\mathfrak{M}_\eta \textit{ if } \lambda \textit{ is limit.}$$

COROLLARY 2.1. *Let L be the dyadic second order language for one ordering,* $\{\mathfrak{A}_\lambda\}$ *the class of its standard models as above. If* κ *is regular and reflects for this L,* $\{\mathfrak{A}_\lambda\}$ *then* $\kappa \in \mathfrak{M}_\kappa$. *Thus if we assume AD, there is some* $\kappa < o({}^R2)$ *such that* $\kappa \in \mathfrak{M}_\kappa$.

Proof. Wellorder the class $\{(\eta,\zeta)\colon \eta,\zeta \textit{ ordinals}\}$ first by the maximum and then lexicographically, let $G(\xi)$ be the ξth pair, let $\langle\eta,\zeta\rangle$ be the mapping inverse to G, so that

$$G(\langle\eta\ \zeta\rangle) = (\eta,\zeta).$$

Clearly G and its inverse are definable in L and G restricted to any regular ordinal ξ gives a bijection of ξ with $\xi \times \xi$.

If A, B are subsets of κ, put

$$A * B = \{\langle\eta,0\rangle\colon \eta \in A\} \cup \{\langle\eta,1\rangle\colon \eta \in B\}.$$

If κ is regular and $(\lambda,(A*B)\cap\lambda) \preccurlyeq\prec (\kappa, A*B)$, then it is immediate that $(\lambda, A\cap\lambda) \preccurlyeq\prec (\kappa, A)$, $(\lambda, B\cap\lambda) \prec\prec (\kappa, B)$. Similarly, if $\xi < \kappa$ and $\{A_\eta\}_{\eta<\xi}$ is a collection of subsets of κ, put

$$*\ \{A_n\}_{n<\xi} = \{\langle\eta,\zeta\rangle\colon \eta < \xi \ \&\ \zeta \in A_\eta\}.$$

Again it is easy to verify that if κ is regular, $\lambda > \xi$ and $(\lambda, *\{A_\eta\}_{\eta<\xi} \cap \lambda) \prec\prec (\kappa, *\{A_\eta\}_{\eta<\xi})$, then for each $\eta < \xi$ $(\lambda, A_\eta \cap \lambda) \prec\prec (\kappa, A_\eta) \preccurlyeq\prec$ as below.

Suppose A is closed unbounded in κ, and $(\lambda, A\cap\lambda) \prec\prec (\kappa, A)$; then $\lambda \in A$, because the sentence "*A is unbounded*" holds in $(\mathfrak{A}_\kappa, A)$, hence it holds in $(\mathfrak{A}_\lambda, A\cap\lambda)$, here λ is a limit of A and since A is closed, $\lambda \in A$.

After these preliminaries we prove, by transfinite induction on $\xi < \kappa$, that $\kappa \in \mathfrak{M}_\xi$.

(1) $\kappa \in \mathfrak{M}_0$, since κ is regular.

(2) $\kappa \in \mathfrak{M}_1$. **Proof.** Given a closed, unbounded A, choose λ so that $(\lambda, A\cap\lambda) \prec\prec (\kappa, A)$. By the remarks above, $\lambda \in A$ and since κ is regular and "*the universe is regular*" is definable in L, λ is also regular.

(3) If $\kappa \in \mathfrak{M}_{\xi+1}$, *then* $\kappa \in \mathfrak{M}_{\xi+2}$. **Proof.** Let A be any closed unbounded subset of κ, write $\mathfrak{M}_\xi$ for $\mathfrak{M}_\xi \cap \kappa$ and choose λ so that

$$(\lambda, (A * \mathfrak{M}_\xi) \cap \lambda) \prec\prec (\kappa, A * \mathfrak{M}_\xi).$$

By the remarks above, $(\lambda, A\cap\lambda) \prec\prec (\kappa, A)$, hence $\lambda \in A$. Also $(\lambda, \mathfrak{M}_\xi \cap \lambda)$

$\prec\!\prec(\kappa, \mathfrak{M}_\xi)$. Since $\kappa \in \mathfrak{M}_{\xi+1}$, the sentence "*every closed unbounded set intersects* $\mathfrak{M}_\xi$" holds in $(\mathfrak{A}_\kappa, \mathfrak{M}_\xi)$, hence it holds in $(\mathfrak{A}_\lambda, \mathfrak{M}_\xi \cap \lambda)$ and $\lambda \in \mathfrak{M}_{\xi+1}$. Thus every closed unbounded $A \subseteq \kappa$ intersects $\mathfrak{M}_{\xi+1}$, hence $\kappa \in \mathfrak{M}_{\xi+2}$.

(4) *If ξ is limit, $\xi < \kappa$, and $\kappa \in \mathfrak{M}_\xi$*, then $\kappa \in \mathfrak{M}_{\xi+1}$.

Proof. Let A be any closed unbounded subset of κ, choose $\lambda > \xi$ so that $(\lambda, B \cap \lambda) \prec\!\prec(\kappa, B)$, where

$$B = * \{\mathfrak{M}_\eta \cap \kappa\}_{\eta<\xi} * A .$$

Now $(\lambda, A \cap \lambda) \prec\!\prec(\kappa, A)$, hence $\lambda \in A$. Also for each $\eta < \xi$, $(\lambda, \mathfrak{M}_\eta \cap \lambda) \prec\!\prec(\kappa, \mathfrak{M}_\eta \cap \kappa)$, and each sentence "*every closed unbounded subset of* κ *intersects* $\mathfrak{M}_b$" is true in $(\mathfrak{A}_\kappa, \mathfrak{M}_\eta \cap \kappa)$, since $\kappa \in \mathfrak{M}_\xi \subseteq \mathfrak{M}_{\eta+1}$, hence each of these sentences holds in $(\mathfrak{A}_\lambda, \mathfrak{M}_\eta \cap \lambda)$, hence $\lambda \in \bigcap_\eta \mathfrak{M}_{\eta+1} = \mathfrak{M}_\xi$. Hence each closed unbounded subset of κ intersects $\mathfrak{M}_\xi$, i.e. $\kappa \in \mathfrak{M}_{\xi+1}$.

§5. The effect of structure properties. Let Γ be a class of subsets of R and to avoid repeating a cumbersome notation, put for this section

$$\Delta = \Gamma \cap \neg\Gamma .$$

We try to find lower bounds for $o(\Delta)$ by assuming that Γ satisfies various structure properties, e.g. closure properties, parametrization, prewellordering, etc.

THEOREM 3. *Assume AD. Let Γ contain all singletons and be closed under continuous substitution, &, $\vee$, $\exists m$, $\forall m$ and either $\exists\alpha$ or $\forall\alpha$. Then $o(\Delta)$ is a cardinal. Moreover, if $\leqq$ is a prewellordeing in Δ with length ξ, then for each $A \subseteq \xi$, $Cod(A, \leqq)$ is in Δ.*

Proof. Take the case when Γ is closed under $\exists\alpha$, the dual case following by applying this to $\neg\Gamma$, suppose $\lambda < \kappa = o(\Delta)$ and $f: \lambda \to \kappa$ is a bijection, let $\leqq$ be a prewellordering in Δ of length λ. Since $\lambda > 0$, we can assume that $Field(\leqq) = R$. Consider the relations on λ

$$P(\eta, \zeta) \Leftrightarrow f(\eta) \leqq f(\zeta),$$

$$Q(\eta, \zeta) \Leftrightarrow f(\eta) < f(\zeta).$$

By Lemma 5 both $Cod(P; \leqq)$, $Cod(Q; \leqq)$ are in $\Sigma^1_1(\leqq)$ so that by the closure properties of Γ both these sets are in Γ. The relation

$$\alpha \leqq' \beta \Leftrightarrow (\alpha, \beta) \in Cod(P; \leqq)$$

is evidently a prewellordering of R with length κ, since if $\phi: R \twoheadrightarrow \lambda$ is the canonical order-preserving surjection for the prewellordering $\leqq$, we have

$$\alpha \leqq' \beta \Leftrightarrow f(\phi(\alpha)) \leqq f(\phi(\beta)).$$

Now $\leqq'$ is in Γ, and also

$$\begin{aligned} \neg(\alpha \leqq' \beta) &\Leftrightarrow f(\phi(\beta)) < f(\phi(\alpha)) \\ &\Leftrightarrow (\beta, \alpha) \in Cod(Q; \leqq), \end{aligned}$$

so that $\leqq'$ is in $\neg\Gamma$, i.e. $\leqq'$ is in Δ and the closure properties of Γ imply that $\kappa + 1$ is also realized in Δ, which is absurd.

The second part of the theorem follows easily from Lemma 5, applied to A and $\xi - A$.

This result already shows that, assuming AD, each $o(\mathbf{\Delta}^1_k)$ is a cardinal. In §7 we shall argue that in $L[R]$ each class $\mathbf{\Sigma}^2_k$ is closed under $\exists m$, $\forall m$, $\exists\alpha$, $\forall\alpha$; under this hypothesis and AD, the theorem also implies that each $o(\mathbf{\Delta}^2_k)$ is a cardinal.

The remaining results of this section will imply that if we assume AD (and other hypotheses, true in $L[R]$), then each $o(\mathbf{\Delta}^2_k)$ and each $o(\mathbf{\Delta}^1_k)$ for odd k are regular. The problem is open for $o(\mathbf{\Delta}^1_k)$ with even $k > 0$.

The *dyadic second order language for two orderings* has two sorts of individual variables, relation symbols $\leqq_1, \leqq_2, =, \in$ and *six* sorts of set variables over subsets of each of the individual domains and over subsets of the four cartesian products determined by the two individual domains. For each pair λ, κ of ordinals, the standard model of this language is

$$\mathfrak{A}_{\lambda,\kappa} = (\lambda, \kappa, {}^{\lambda}2, {}^{\kappa}2^{(\lambda\times\lambda)}2, {}^{(\lambda\times\kappa)}2, {}^{(\kappa\times\lambda)}2, {}^{(\kappa\times\kappa)}2, \leqq_\lambda, \leqq_\kappa, \varepsilon),$$

where $\leqq_\lambda$, $\leqq_\kappa$ are the orderings on λ and κ respectively.

In §3 we extended the language of analysis by prime formulas of the form $\chi(\delta) \simeq m$ in order to define $\Sigma^1_1(\chi)$. It is clear how we can extend the language of analysis by relation symbols $P^*_1, \ldots, P^*_m$, each P^*_i suitable for denoting a subset of some product space $\mathfrak{X}_i$. We call formulas of this extended language *analytic* in $P^*_1, \ldots, P^*_m$. A relation definable by a formula analytic in $P^*_1, \ldots, P^*_m$ when we give specific values $P_1, \ldots, P_m$ to the symbols $P^*_1, \ldots, P^*_m$ *is analytic in* $P_1, \ldots, P_m$ (*projective in* $P_1, \ldots, P_m$ if we also allow constants from R as parameters).

Let $\leqq_1, \leqq_2$ be prewellorderings of subsets of R, let $\phi: Field(\leqq_1) \twoheadrightarrow \lambda$, $\psi: Field(\leqq_2) \twoheadrightarrow \kappa$ be the canonical surjections. To each subset $A \subseteq \lambda \times \kappa$, put

$$Cod(A; \leqq_1, \leqq_2) = \{(\alpha, \beta): \alpha \in Field(\leqq_1) \;\&\; \beta \in Field(\leqq_2) \;\&\; (\phi(\alpha), \psi(\beta)) \in A\}$$

and similarly for subsets of $\kappa \times \lambda$. In order to have a uniform notation, let us also put

$$Cod(A; \leqq_1, \leqq_2) = Cod(A; \leqq_1),$$

$$Cod(B; \leqq_1, \leqq_2) = Cod(B; \leqq_2),$$

for subsets $A \subseteq \lambda$, $B \subseteq \kappa$, at least for the next lemma.

LEMMA 8. *Assume AD. To each formula* $\theta(x_1, \dots, x_n, y_1, \dots, y_m)$ *of the dyadic second order language for two orderings with the exhibited individual variables and set variables* $A_1, A_2, \dots$ *we can assign a formula* $\theta^*(\alpha_1, \dots, \alpha_n, \beta_1, \dots, \beta_m)$, *analytic in the symbols* $\leqq_1^*, \leqq_2^*, A_1^*, A_2^*, \dots$ *with the following property. If* $\leqq_1, \leqq_2$ *are prewellorderings of subsets of* R *with lengths* λ, κ *and canonical surjections* ϕ: $Field(\leqq_1) \twoheadrightarrow \lambda$, ψ: $Field(\leqq_2) \twoheadrightarrow \kappa$, *if* $B_1, B_2, \dots$ *are values for the set variables* $A_1, A_2, \dots$ *in the standard model* $\mathfrak{A}_{\lambda,\kappa}$ *and if we interpret* $\leqq_1^*, \leqq_2^*, A_1^*, A_2^*, \dots$ *by* $Cod(B_1; \leqq_1, \leqq_2)$, $Cod(B_2; \leqq_1, \leqq_2)$, ..., *respectively, then for* $\alpha_1, \dots, \alpha_n \in Field(\leqq_1)$, $\beta_1, \dots, \beta_m \in Field(\leqq_2)$ *we have*

$$\theta^*(\alpha_1, \dots, \alpha_n, \beta_1, \dots, \beta_m) \Leftrightarrow \mathfrak{A}_{\lambda,\kappa} \models \theta(\phi(\alpha_1), \dots, \phi(\alpha_n), \psi(\beta_1), \dots, \psi(\beta_m)).$$

Proof is by induction on the construction of the formulas $\theta(x_1, \dots, x_n, y_1, \dots, y_m) = \theta$. The assignment of θ^* to θ for prime θ is trivial, e.g. to $x_1 \leqq_1 x_2$ we assign $\alpha_1 \leqq_1^* \alpha_2$, to $y_1 = y_2$ we assign $\beta_1 \leqq_2^* \beta_2$ & $\beta_2 \leqq_2^* \beta_1$ to $(x_1, y_1) \in A_1$ we assign $(\alpha_1, \beta_1) \in A_1$, etc. The induction steps involving the propositional connectives and quantifiers over λ or κ are also trivial, e.g. to $\exists x \theta(x)$ we assign $\exists \alpha[\alpha \leqq_1 \alpha \;\&\; \theta^*(\alpha)]$. It only remains to show how we deal with the set quantifiers.

Given $\leqq_1, \leqq_2$, let $\leqq$ be the lexicographical prewellordering on the cartesian product of the fields,

$$(\alpha, \beta) \leqq (\alpha', \beta') \Leftrightarrow \alpha, \alpha' \in Field(\leqq_1) \;\&\; \beta, \beta' \in Field(\leqq_2)$$
$$\& \{[\alpha \leqq_1 \alpha' \;\&\; \neg(\alpha' \leqq_1 \alpha')] \vee [\alpha \leqq_1 \alpha' \;\&\; \alpha' \leqq_1 \alpha \;\&\; \beta \leqq_2 \beta']\}.$$

The length of $\leqq$ is precisely $\kappa \cdot \lambda$, i.e. the order type of $\lambda \times \kappa$ ordered by the lexicographical wellordering. If χ: $Field(\leqq) \twoheadrightarrow \kappa \cdot \lambda$ is the canonical surjection and $f: \lambda \times \kappa \twoheadrightarrow \kappa \cdot \lambda$ the unique similarity, it is immediate that the diagram

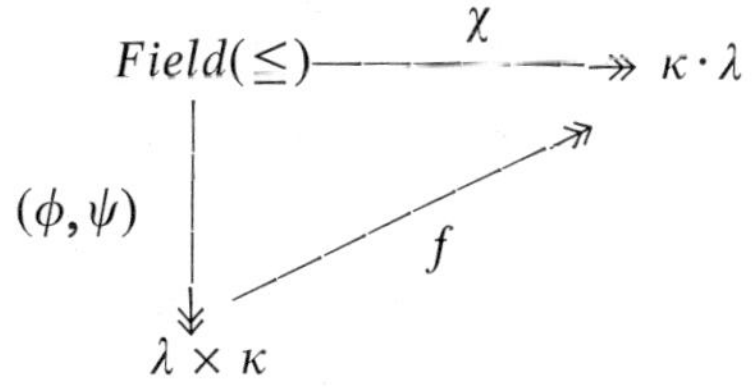

commutes where $(\phi, \psi)(\alpha, \beta) = (\phi(\alpha), \psi(\beta))$. From this follows that for every subset $B \subseteq \lambda \times \kappa$,

$$Cod(f[B]; \leqq) = \{(\alpha, \beta) : \alpha \in Field(\leqq_1) \;\&\; \beta \in Field(\leqq_2) \;\&\; (\phi(\alpha), \psi(\beta)) \in B\}$$
$$= Cod(B; \leqq_1, \leqq_2).$$

Thus by Lemma 5, for each subset $B \subseteq \lambda \times \kappa$, $Cod(B; \leqq_1, \leqq_2)$ *is* $\Sigma_1^1(\leqq)$ as a subset of $R \times R$. Now $G(\leqq)$ is analytic in $\leqq_1, \leqq_2$, by a formula that we can construct from (3–3). Moreover the relation

$$(5\text{–}1) \qquad Q(\beta) \Leftrightarrow (\exists B \subseteq \lambda \times \kappa)[G_\beta(\leqq) = Cod(B; \leqq_1, \leqq_2)]$$

is also analytic in $\leqq_1, \leqq_2$, since it is equivalent to

$$(5\text{–}2) \qquad (\forall \gamma, \gamma', \delta, \delta')[(\beta, \gamma, \delta) \in G(\leqq) \;\&\; \gamma \leqq_1 \gamma' \;\&\; \gamma' \leqq_1 \gamma \;\&\; \delta \leqq_2 \delta' \;\&\; \delta' \leqq_2 \delta \Rightarrow (\beta, \gamma', \delta') \in G(\leqq)].$$

Thus if $\theta^*(A^*)$ is assigned to $\theta(A)$, with A a variable over subsets of $\lambda \times \kappa$, we assign to $\exists A \theta(A)$

$$\exists \beta[Q^*(\beta) \;\&\; \theta^*(G_\beta(\leqq))],$$

where $Q^*(\beta)$ is the formal version of $Q(B)$. The quantifiers over subsets of the other cross products of λ and κ or λ and κ themselves are treated similarly, so that the proof is complete.

This lemma embodies many of the useful consequences of Lemma 5. It is obvious that it holds for the full second order language for any number of orderings.

THEOREM 4. *Assume AD. Let Λ be a class of sets containing all singletons and closed under continuous substitution, &, $\neg$, $\exists m$, $\exists \alpha$. Then $o(\Lambda)$ is a limit cardinal. Moreover, if $\leqq$ is a prewellordering in Λ with length ξ, then for each $A \subseteq \xi$, $Cod(A; \leqq)$ is in Λ.*

Proof. To prove that $o(\Lambda)$ is a limit cardinal, let λ be the length of some prewellordering $\leqq$ in Λ, we shall show that λ^+ is also realized in Λ. In the dyadic second order language for one ordering (a sublanguage of that for two orderings) with variables $A_1, A_2, \ldots$ varying over subsets of $\lambda \times \lambda$ in the standard model $\mathfrak{A}_\lambda$, let $\theta(A_1, A_2)$ be the formal version of

> "*A_1, A_2 are wellorderings of λ and there is some subset of $\lambda \times \lambda$ which is an isomorphism of A_1 onto A_2 or onto an initial segment of A_2*".

Let $\theta^*(A_1^*, A_2^*)$ be the formula analytic in $\leqq^*, A_1^*, A_2^*$ assigned to $\theta(A_1, A_2)$ by Lemma 8, let $G(\leqq)$ be universal for the $\Sigma_1^1(\leqq)$ subsets of $R \times R$. Clearly G is analytic in $\leqq$, hence in Λ. As in the proof of Lemma 8, the relation

$$Q(\alpha) \Leftrightarrow (\exists A \subseteq \lambda \times \lambda)[G_\alpha(\leqq) = Cod(A; \leqq)]$$

is analytic in $\leqq$, hence in Λ. Put then

$$\alpha \leqq' \beta \Leftrightarrow Q(\alpha) \;\&\; Q(\beta) \;\&\; \theta^*(G_\alpha(\leqq), G_\beta(\leqq)).$$

Now $\leqq'$ is in Λ, and it is a prewellordering with length λ^+.

The second assertion of the theorem is immediate from Lemma 5.

THEOREM 5. *Assume AD. Let Γ be parametrized, containing all singletons and closed under continuous substitution, &, $\vee$, $\exists m$, $\forall m$, $\exists\alpha$ and $\forall\alpha$. Then $o(\Delta)$ is a regular limit cardinal (inaccessible). Moreover, if $\leqq$ is a prewellordering in Δ with length ξ and $A \subseteq \xi$, then $Cod(A; \leqq)$ is in Δ.*

Proof. Let $\kappa = o(\Delta)$, the preceding theorem implies that κ is a limit cardinal and the second assertion of the theorem follows from Lemma 5. It remains to show that κ is regular.

Let λ be the length of some prewellordering $\leqq$ of R in Δ and assume that $f: \lambda \to \kappa$ is a cofinal map — we shall obtain a contradiction. We may assume that for each η, $f(\eta) > 0$. Let G be universal for the subsets of $R \times R$ in Γ, put

$$g(\eta) = \{(\beta, \gamma) \colon G_\beta = R \times R - G_\gamma \textit{ and } G_\beta \textit{ is a prewellordering of } R \textit{ with length } f(\eta)\}.$$

By hypothesis $g(\eta) \neq \emptyset$ for each η, so by Lemma 3 there is a choice subfunction g^* of g such that $Cod(g^*, \leqq)$ is in $\Sigma^1_1(\leqq)$, hence by the closure properties of Δ, $Cod(g^*, \leqq) \in \Delta$.

If $(\alpha, \beta, \gamma) \in Cod(g^*; \leqq)$, then G is a prewellordering of R with length $f(\phi(\alpha))$, where $\phi: R \twoheadrightarrow \lambda$ is the canonical order preserving surjection for $\leqq$; let $\phi_\beta \colon R \twoheadrightarrow f(\phi(\alpha))$ be the canonical order preserving surjection for G_β. Put

$$\begin{aligned}(\alpha, \beta, \gamma, \delta) \leqq' (\alpha', \beta, \gamma', \delta') \Leftrightarrow\ & (\alpha, \beta, \gamma) \in Cod(G^*; \leqq) \\ & \&\ (\alpha', \beta', \gamma') \in Cod(g^*; \leqq) \\ & \&\ \phi_\beta(\delta) \leqq \phi_{\beta'}(\delta').\end{aligned}$$

Now $\leqq'$ is a prewellordering of a subset of R^4 with length precisely κ, the canonical surjection being

$$\psi(\alpha, \beta, \gamma, \delta)) = \phi_\beta(\delta).$$

We shall obtain a contradiction and complete the proof if we can show that $\leqq'$ is in Δ.

In the dyadic second order language for two orderings, with $x, x', \ldots$ varying over the first ordering and $y, y', \ldots$ over the second, let $\phi(x, y)$ be the formula which defines the relation $x \leqq y$, i.e. *there exists a subset of the cross product of the two domains which is an order preserving bijection of $\{x' : x' \leqq_1 x\}$ onto $\{y' : y' \leqq_2 y\}$*. Let $\theta^*(\alpha, \beta, \leqq^*_1, \leqq^*_2)$ be the analytic in $\leqq_1^*, \leqq_2^*$ formula associated with $\theta(x,y)$ by Lemma 8. If $(\alpha,\beta,\gamma,\delta)$, $(\alpha',\beta',\gamma',\delta')$ are in the field of $\leqq'$, then Lemma 8 asserts precisely that

$$\phi_\beta(\delta) \leqq \phi_{\beta'}(\delta') \Leftrightarrow \theta^*(\delta,\delta',\leqq_1^*,\leqq_2^*)$$

whenever $\leqq_1^*$ interprets the relation G_β and $\leqq_2$ the relation $G_{\beta'}$. But $G_\beta = R \times R - G_\gamma$, $G_\beta' = R \times R - G_{\gamma'}$; if we substitute $G_\beta(G_{\beta'})$ in all the positive occurrences of $\leqq_1^*$ $(\leqq_2^*)$ and $R \times R - G_\gamma$ $(R \times R - G_{\gamma'})$ in all the negative occurrences of $\leqq_1^*$ $(\leqq_2^*)$, then the closure properties of Γ imply that the resulting relation is in Γ. The dual substitution gives us a relation in $\neg\Gamma$, and both these relations are equivalent to $\phi_\beta(\delta) \leqq \phi_{\beta'}(\delta')$ whenever $(\alpha,\beta,\gamma,\delta), (\alpha',\beta',\gamma',\delta')$ are in the field of $\leqq'$, i.e. $(\alpha,\beta,\gamma),(\alpha'\beta',\gamma') \in Cod(g^*;\leqq)$. Thus $\leqq'$ is in Δ and the proof that κ is regular is complete.

If we want to eliminate the hypothesis that Γ is closed under *both* $\exists\alpha$ and $\forall\alpha$, we must add the prewellordering property to the assumptions.

THEOREM 6. *Assume AD. Let Γ be parametrized, containing all singletons and closed under continuous substitution, &, $\vee$, $\exists m$, $\forall\alpha$, assume Prewellordering (Γ). Then $o(\Delta)$ is regular. Moreover, if $\leqq$ is a prewellordering of the universal set G with length κ which establishes the prewellordering property, then $\kappa = o(\Delta)$ and for each $A \subseteq \kappa$, $Cod(A;\leqq)$ is in Γ.*

Proof. The statement of the prewellordering property implies immediately that $\kappa \leqq o(\Delta)$, since for each $\lambda < \kappa$, if $\phi: G \twoheadrightarrow \kappa$ is the canonical surjection and $\phi(x_0) = \lambda$, then

$$\{(y,z)\colon y \leqq z \;\&\; z \leqq x_0 \;\&\; \neg(x_0 \leqq z)\} \tag{5–3}$$

is in Δ and has length λ.

Assume $\leqq'$ is a prewellordering in Δ with length λ and canonical surjection ψ: *Field*$(\leqq) \twoheadrightarrow \lambda$, assume that there is some $f: \lambda \to \kappa$ such that *supremum*$\{f(\eta)\colon \eta < \lambda\} = \kappa$. We shall obtain a contradiction from these assumptions, thereby showing at the same time that $o(\Delta) \leqq \kappa$ and that κ is regular. Put

$$f^*(\eta) = \{x\colon x \in G \;\&\; \phi(x) = f(\eta)\}$$

and by Lemma 3 let g^* be a choice subfunction of f^* with $Cod(g^*;\leqq')$ in $\Sigma_1^1(\leqq')$. The closure properties of Γ imply that $Cod(g^*;\leqq')$ is in $\neg\Gamma$. Now

$$y \in G \Leftrightarrow \exists\alpha\exists x[(\alpha,x) \in Cod(g^*;\leqq') \;\&\; y \mathrel{\dot{\leqq}} x],$$

where $\dot{\leqq}$ is in $\neg\Gamma$ and agrees with $\leqq$ when its second argument is in G

as in (2–2), which implies $G \in \neg\Gamma$, violating the remark at the end of 2.1 (the "hierarchy theorem").

It remains to show that for each $A \subseteq \kappa$, $Cod(A; \leqq)$ is in Γ — notice that this does not follow immediately from Lemma 5, since Γ is not assumed closed under $\exists\alpha$.

For each $\lambda < \kappa$, let $\leqq_\lambda$ be the initial segment of $\leqq$ up to and not including points that correspond to λ, defined by (5–3). For each $A \subseteq \lambda$, $Cod(A; \leqq_\lambda)$ is in $\Sigma^1_1(\leqq_\lambda)$, hence in $\neg\Gamma$ by the closure properties of Γ; but this implies that for each $A \subseteq \lambda$, $Cod(A; \leqq_\lambda)$ is in Γ, since

$$x \in Cod(A; \leqq_\lambda) \Leftrightarrow x \leqq_\lambda x \ \&\ x \notin Cod(\lambda - A; \leqq_\lambda).$$

Let H be universal for the subsets of $R \times R$ in Γ, let $\pi : R \to R \times R$ be the canonical homeomorphism, let $A \subseteq \kappa$, consider the following two-person game. If I plays α and II plays β,

$$\text{II } wins \Leftrightarrow \pi\alpha \notin G \ \vee\ [\pi\alpha \in G \ \&\ (\exists\lambda > \phi(\alpha))[H_\beta = Cod(A \cap \lambda; \leqq_\lambda)]].$$

(Here of course $\phi : G \twoheadrightarrow \kappa$ is the canonical surjection.)

Suppose I wins the game with a strategy σ, we shall obtain a contradiction. We must have for all β, $\pi(\sigma * [\beta]) \in G$ and $supremum\{\phi(\pi(\sigma * [\beta])) : \beta \in R\} = \kappa$, since if for all β, $\phi(\pi(\sigma * [\beta])) < \lambda$, then II could win by playing β so that $H_\beta = Cod(A \cap \lambda; \leqq_\lambda)$. Therefore we have

$$y \in G \Leftrightarrow \exists\beta[y \stackrel{\cdot}{\leqq} \pi(\sigma * [\beta])]$$

which implies that $G \in \neg\Gamma$, a contradiction.

Now II must win, by some strategy τ, and we have

$$x \in Cod(A; \leqq) \Leftrightarrow x \in H_{[\pi^{-1}x] * \tau}$$

$$\Leftrightarrow ([\pi^{-1}x] * \tau, x) \in H$$

which shows that $Cod(A; x) \in \Gamma$, completing the proof.

If we assume that Γ satisfies the closure properties of Theorem 5 *uniformly*, in a sense we shall explain shortly, then we can prove $\kappa = o(\Delta)$ without assuming AD. This result (and the lemmas leading to it) has independent interest in hierarchy theory and is the best characterization of $o(\Delta)$ that we know, without assuming axioms that contradict choice. For these reasons we give it here, even though it is not directly related to AD.

If Γ is parametrized by G, then $\Gamma = \{G_\alpha : \alpha \in R\}$ and for each $A \in \Gamma$, we call $\{\alpha : G_\alpha = A\}$ the set of *codes* for A. In (2–1) we assigned to each parametrization of Γ a canonical parametrization of $[\mathfrak{X}]\Gamma$ (when Γ is closed under recursive substitution) so that we can talk about codes for

$A \in \Gamma$ when $A \subseteq \mathfrak{X}$, for any product space $\mathfrak{X}$. Our standard hypotheses up till now are that Γ is closed under various operations, e.g. continuous substitution, &, $\vee$, etc. Let us say that Γ is *uniformly closed* under an operation if the result of applying that operation is always in Γ and a code of it can be computed from codes of the arguments via a *continuous* function. To be precise for some of the operations that we shall be most concerned with: Γ is uniformly closed under continuous preimages if there exists a continuous $f: R \times R \to R$ such that if $\{\varepsilon\}^{R;R}$ is completely defined, then $G_{f(\varepsilon,\alpha)} = \{\beta : \{\varepsilon\}(\beta) \in G_\alpha\}$. Γ is uniformly closed under &, if there is a continuous $f: R \times R \to R$ such that for each α, β, $G_{f(\alpha,\beta)} = G_\alpha \cap G_\beta$, and similarly for $\vee$. Γ is uniformly closed under $\exists m$ if there is a continuous function $f: R \to R$ such that for each α, $G_{f(\alpha)} = \{\beta : \exists m(m,\beta) \in H_\alpha\}$, where H is the canonical universal set for $[\omega \times R]\Gamma$, defined from G via (2–1), and similarly for $\forall m$, $\exists \alpha$, $\forall \alpha$.

LEMMA 9. *Let Γ be parametrized by G, containing all singletons and uniformly closed under continuous substitution, &, $\vee$, $\exists m$, $\forall m$, $\forall \alpha$, assume Prewellordering (Γ) via a canonical surjection $\phi: G \twoheadrightarrow \kappa$. There is a continuous function $f_1: R \times R \to R$ such that if $A = R - G_\alpha$ and if for each $\beta \in A$, $\{\varepsilon\}(\beta) \in G$, then $f(\varepsilon, \alpha) \in G$ and $(\forall \beta \in A)[\phi(\{\varepsilon\}(\beta)) < \phi(f(\varepsilon,\alpha))]$.*

The import of the lemma is that if a continuous partial function is defined over some $A \in \neg\Gamma$ and takes values in G, then the set of these values is bounded in the prewellordering. We need the *uniform version* of this lemma, however, in the way that we stated it.

Proof. Put

$$\beta \in B \Leftrightarrow \exists \gamma[\gamma \in A \;\&\; \{\varepsilon\}(\gamma) \text{ is defined } \&(\beta,\beta) \dot{\leqq} \{\varepsilon\}(\gamma)]$$

where $\dot{\leqq}$ is in $\neg\Gamma$ and agrees with $\leqq$ when its second argument is in G, as in (2–2). Now the closure properties of Γ imply that $B \in \neg\Gamma$; moreover, because of the uniformity hypothesis, there is a continuous $g_1: R \times R \to R$ such that if $A = R - G_\alpha$, then $B = R - G_{g_1(\varepsilon,\alpha)}$, i.e.

$$\beta \in B \text{ s}\Leftrightarrow (g_1(\varepsilon,\alpha),\beta) \notin G$$

Put

$$f_1(\varepsilon,\alpha) = (g_1(\varepsilon,\alpha), g_1(\varepsilon,\alpha))$$

and assume $A = R - G_\alpha$ and that for each $\beta \in A$, $\{\varepsilon\}(\beta) \in G$. If $f_1(\varepsilon,\alpha) \notin G$, then by definition $g_1(\varepsilon,\alpha) \in B$, hence for some $\gamma \in A$, $(g_1(\varepsilon,\alpha), g_1(\varepsilon,\alpha)) = f_1(\varepsilon,\alpha) \dot{\leqq} \{\varepsilon\}(\gamma)$, which implies $f_1(\varepsilon,\alpha) \in G$, contradiction; hence $f_1(\varepsilon,\alpha) \in G$. But then $g_1(\varepsilon,\alpha) \notin B$, which implies that for each $\gamma \in A$, $\neg(f_1(\varepsilon,\alpha) \dot{\leqq} \varepsilon\}(\gamma))$, i.e. $\neg(f_1(\varepsilon,\alpha) \leqq \{\varepsilon\}(\gamma))$, i.e. $\phi(\{\varepsilon\}(\gamma)) < \phi(f_1(\varepsilon,\alpha))$.

LEMMA 10. *Let Γ be parametrized by G, containing all singletons and uniformly closed under continuous substitution, &, $\vee$, $\exists m$, $\forall m$, $\forall\alpha$, assume Prewellordering* (Γ) *via a prewellordering $\leqq$ on G with length κ and canonical surjection $\phi: G \twoheadrightarrow \kappa$. There is a continuous function $f_2(\alpha)$ such that if H is the universal set for $[R \times R]\Gamma$ associated with G via* (2–1) *and $\leqq' = H_\alpha = R \times R - H_\beta$ is a prewellordering with length ξ, then $f_2(\alpha, \beta) \in G$ and $\xi \leqq \phi(f_2(\alpha, \beta))$.*

Again, the import of the lemma is that every prewellordering in Δ has length less than κ, but it is this stronger, uniform version that our proof is suited for.

Proof. Using the uniform closure properties of Γ we easily show that for some continuous $g_1(\alpha, \beta, \gamma)$, if $\leqq' = H_\alpha = R \times R - H_\beta$ is some relation in Δ, and $\gamma \leqq' \gamma$ is some element of its field, then

$$R - G_{g_1(\alpha,\beta,\gamma)} = \{\delta : \delta \leqq' \gamma \mathbin{\&} \neg(\gamma \leqq' \delta)\}.$$

Using the recursion theorem for continuous partial functions, choose ε^* so that

$$\{\varepsilon^*\}(\alpha, \beta, \gamma) \simeq f_1(g_1(\alpha, \beta, \gamma), S^{R\times R}(\varepsilon^*, \alpha, \beta)), \tag{5–4}$$

where f_1 is the function of the preceding lemma and $S^{R\times R}$ is the recursive function of the iteration theorem, so that

$$\{\varepsilon^*\}(\alpha, \beta, \gamma) \simeq \{S^{R\times R}(\varepsilon^*, \alpha, \beta)\}(\gamma). \tag{5–5}$$

Notice that $\{\varepsilon^*\}(\alpha, \beta, \gamma)$ is a completely defined function, since f_1 and g_1 are completely defined.

Assume that $\leqq' = H_\alpha = R \times R - H_\beta$ is a prewellordering with length ξ and canonical surjection $\psi: \mathit{Field}(\leqq') \twoheadrightarrow \xi$. We prove by transfinite induction on $\psi(\gamma)$ that if $\gamma \in \mathit{Field}(\leqq')$, then $\{\varepsilon^*\}(\alpha, \beta, \gamma) \in G$ and $\psi(\gamma) \leqq \phi(\{\varepsilon^*\}(\alpha, \beta, \gamma))$: assuming this for all δ with $\psi(\delta) < \psi(\gamma)$, the preceding lemma, (5–4) and (5–5) give us immediately that for all δ with $\psi(\delta) < \psi(\gamma)$, we have $\phi(\{\varepsilon^*\}(\alpha, \beta, \delta)) < \phi(\{\varepsilon^*\}(\alpha, \beta, \gamma))$, i.e. $\psi(\delta) < \phi(\{\varepsilon^*\}(\alpha, \beta, \gamma))$, hence $\psi(\gamma) \leqq \phi(\{\varepsilon^*\}(\alpha, \beta, \gamma))$.

The lemma now follows if we choose by the uniform closure a continuous $g_2: R \times R \to R$ such that if $\leqq' = H_\alpha = R \times R - H_\beta$, then

$$R - G_{g_2(\alpha,\beta)} = \{\gamma : \gamma \leqq' \gamma\}$$

and put

$$f_2(\alpha, \beta) = f_1(g_2(\alpha, \beta), S^{R\times R}(\varepsilon^*, \alpha, \beta)).$$

THEOREM 7. *Let Γ be parametrized by G, containing all singletons and uniformly closed under continuous substitution, &, $\vee$, $\exists m$, $\forall m$, $\forall\alpha$,* *assume Prewellordering* (Γ) *via a prewellordering* $\leqq$ *on G with length* κ. *Then* $\kappa = o(\Delta)$.

Proof. That $\kappa \leqq o(\Delta)$ follows from the statement of the prewellordering property, as in the proof of Theorem 5. That $o(\Delta) \leqq \kappa$ follows immediately from the preceding lemma.

§6. Projective ordinals. For each $k > 0$, let

$$\boldsymbol{\delta}_k^1 = o(\boldsymbol{\Delta}_k^1),$$

We collect in one theorem the results about these ordinals which follow from what we have already shown.

THEOREM 8.

(8.1) *Assume Determinacy* $(\boldsymbol{\Delta}_2^1 m)$, *DC. Let* $\leqq$ *be a prewellordering of of a universal* $\boldsymbol{\Pi}_{2m+1}^1$ *set which has length* κ *and establishes Prewellordering* $(\boldsymbol{\Pi}_{2m+1}^1)$; *then* $\kappa = \boldsymbol{\delta}_{2m+1}^1$.

(8.2) *Assume AD.*

(8.2.1) *Each* $\boldsymbol{\delta}_k^1$ *is a cardinal.*

(8.2.2) *If* $\leqq$ *is a projective prewellordering with length* ξ *and* $A \subseteq \xi$, *then Cod*$(A; \leqq)$ *is a projective set, hence A is definable (in set theory) from a real parameter.*

(8.2.3) *If* $\xi < \boldsymbol{\delta}_k^1$ *and* $A = \bigcup_{\eta<\xi} A_\eta$ *with each* $A_\eta \in \boldsymbol{\Sigma}_k^1$, *then* $A \in \boldsymbol{\Sigma}_k^1$.

(8.2.4) *A set A is in* $\boldsymbol{\Sigma}_2^1$ *if and only if it is the union of* $\aleph_1$ *Borel sets.*

(8.3) *Assume AD, DC.*

(8.3.1) *Each* $\boldsymbol{\delta}_{2m+1}^1$ *is a regular cardinal.*

(8.3.2) *If* $\leqq$ *is a prewellordering of a universal* $\boldsymbol{\Pi}_{2m+1}^1$ *set which has length* $\boldsymbol{\delta}_{2m+1}^1$ *and establishes Prewellordering* $(\boldsymbol{\Pi}_{2m+1}^1)$, *then for each* $A \subseteq \boldsymbol{\delta}_{2m+1}^1$, *Cod*$(A; \leqq)$ *is* $\boldsymbol{\Pi}_{2m+1}^1$. *(For* $m = 0$ *this is due to Solovay.)*

(8.3.3) *For each* $m \geqq 1$, $\boldsymbol{\delta}_{2m-1}^1 < \boldsymbol{\delta}_{2m}^1 \leqq \boldsymbol{\delta}_{2m+1}^1$.

(8.3.4) *For each* $m \geqq 1$, $(\boldsymbol{\delta}_{2m-1}^1)^+ < \boldsymbol{\delta}_{2m+1}^1$.

(8.3.5) *For each k,* $\aleph_k^1 \leqq \boldsymbol{\delta}_k^1$.

Proof. (8.1) follows from Theorem 7 and the Prewellordering Theorem, (IV), of §2. (8.2.1), (8.2.2) follow from Theorem 3, (8.2.3) from Lemma 6 and (8.2.4) from (8.2.3) and the classical result that every $\boldsymbol{\Sigma}_2^1$ set is the union of $\aleph_1$ Borel sets. (8.3.1) and (8.3.2) follow from Theorem 6 and (IV) of §2.

The first half of (8.3.3) follows from Theorem 6 and (IV) of §2 (which imply that $\boldsymbol{\delta}_{2m-1}^1$ is the length of a $\boldsymbol{\Pi}_{2m-1}^1$, hence $\boldsymbol{\Delta}_{2m}^1$ prewellordering) and the second half is trivial. (8.3.5) follows immediately from (8.3.3) and (8.3.4) so that we need only give a

Proof of (8.3.4). Let $\leqq^{2m-1}$ be a prewellordering on a set G^{2m-1}, universal for Π^1_{2m-1}, which establishes *Prewellordering* ($\mathbf{\Pi}^1_{2m-1}$) by (IV) of §2. By Theorem 6, each subset A of $\mathbf{\delta}^1_{2m-1}$ has $Cod(A; \leqq^{2m-1})$ in $\mathbf{\Pi}^1_{2m-1}$, and the same holds by a trivial extension for every binary relation A on $\mathbf{\delta}^1_{2m-1}$. If H^{2m-1} is universal for the sets of pairs in $\mathbf{\Pi}^1_{2m-1}$, it is now trivial to verify that the set

$$\alpha \in W^{2m-1} \Leftrightarrow H^{2m-1}_\alpha = Cod(A; \leqq) \textit{ for some wellordering } A \textit{ of } \mathbf{\delta}^1_{2m1}$$

is in $\mathbf{\Pi}^1_{2m}$. Let G^{2m+1} be universal for $\mathbf{\Pi}^1_{2m+1}$, let $\phi^{2m+1}: G^{2m+1} \twoheadrightarrow \mathbf{\delta}^1_{2m+1}$ be the canonical surjection for a prewellordering that establishes *prewellordering* ($\mathbf{\Pi}^1_{2m+1}$). Using the trivial observation that each $\mathbf{\Pi}^1_{2m}$ set is "uniformly" in $\mathbf{\Delta}^1_{2m+1}$ and Lemma 10, we easily obtain a continuous function $f(\alpha)$ such that

$$\alpha \in W^{2m-1} \Rightarrow f(\alpha) \in G^{2m+1}$$

and if $H^{2m-1}_\alpha = Cod(A; \leqq^{2m-1})$ for some wellordering A with length ξ_A, then $\xi_A \leqq \phi^{2m+1}(f(\alpha))$. Now Lemma 9 implies that for some $\eta < \mathbf{\delta}^1_{2m+1}$ and all such A, $\xi_A \leqq \eta$, so that $(\mathbf{\delta}^1_{2m-1})^+ \leqq \eta < \mathbf{\delta}^1_{2m+1}$.

It is a classical result that $\mathbf{\delta}^1_0 = \mathbf{\delta}^1_1 = \aleph_1$. In view of (8.3.5) above one would hope that with *AD* and *DC* either $\mathbf{\delta}^1_k = \aleph^1_k$ for all $k \geqq 1$ or $\mathbf{\delta}^1_{2m} = \mathbf{\delta}^1_{2m+1} = \aleph^1_{2m+1}$ for all m.

It would be nice to be able to prove from *AD* and *DC* that every δ^1_k is regular. However we only know this for odd k and it is not clear that our methods can be extended easily to prove it for even k.

Solovay has proved that $\aleph_1$ and $\aleph_2$ are measurable assuming *AD*. It should be the case that with *AD* and *DC* all $\mathbf{\delta}^1_{2m+1}$ are measurable, perhaps also all $\mathbf{\delta}^1_{2m}$. (It can be easily shown that each $\mathbf{\delta}^1_{2m+1}$ carries a countably additive complete measure such that the measure of every bounded set is 0.)

§7. $\mathbf{\Delta}^2_k$. For each $k \geqq 1$, put

$$\mathbf{\delta}^2_k = o(\mathbf{\Delta}^2_k).$$

The key to obtaining results about $\mathbf{\delta}^2_k$ with our methods is to prove closure properties about the classes $\mathbf{\Sigma}^2_k$, $\mathbf{\Pi}^2_k$, $\mathbf{\Delta}^2_k$. It is easy to prove from the definitions, Lemma 1 and trivial contractions of variables that $\mathbf{\Sigma}^2_k$, $\mathbf{\Pi}^2_k$ are parametrized and closed under continuous substitution, & and $\vee$. However the usual proofs that these classes are also closed under $\exists m$ $\forall m$, $\exists\alpha$, $\forall\alpha$ use the following special case of the axiom of choice.

For each $F: R \to \omega$ put

$$F_\alpha(\beta) = F(\pi(\alpha, \beta)),$$

where $\pi\colon R \times R \to R$ is the canonical recursive homeomorphism. The axiom of choice then implies that for each $A \subseteq R \times {}^R\omega$

$$\text{(7–1)} \qquad \forall\alpha\exists F(\alpha, F) \in A \Rightarrow \exists F\forall\alpha(\alpha, F_\alpha) \in A$$

and (7–1) then implies easily that $\mathbf{\Sigma}_k^2$, $\mathbf{\Pi}_k^2$ are closed under $\exists m$, $\forall m$, $\exists\alpha$, $\forall\alpha$. We do not know if (7–1) is consistent with AD; but the same closure properties can be proved from the more innocuous *collection* property, which we can establish at least in $L[R]$,

Collection. For each $A \subseteq R \times {}^R\omega$,

$$\forall\alpha\exists F(\alpha, F) \in A \Rightarrow \exists F\forall\alpha\exists\beta(\alpha, F_\beta) \in A.$$

LEMMA 11. *Let M be a class which is a model of ZF (without choice), which contains R and such that some surjection $f\colon ON \times R \twoheadrightarrow M$ is definable in M with parameters from M; then M satisfies Collection. In particular $L[R]$ and the class of sets hereditarily ordinal definable from real numbers satisfy Collection.*

Proof. Assume $V = M$ and $\forall\alpha\exists F(\alpha, F) \in A$, for some $A \subseteq R \times {}^R\omega$. Put

$$\eta(\alpha) = \mathit{infimum}\{\xi : \exists\gamma[f(\xi, \gamma) \in {}^R\omega \ \&\ (\alpha, f(\xi, \gamma)) \in A]\},$$

$$G(\alpha, \gamma, \delta) = f(\eta(\alpha), \gamma)(\delta) \quad \text{if} \quad f(\eta(\alpha), \gamma) \in {}^R\omega,$$

$$= 0 \qquad \text{otherwise,}$$

and finally

$$F(\varepsilon) = G(\alpha, \gamma, \delta)$$

for the unique α, β, γ, δ such that $\pi(\beta, \delta) = \varepsilon$ and $\pi(\alpha, \gamma) = \beta$. Now for each α, choose γ so that $f(\eta(\alpha), \gamma) \in {}^R\omega$ and $(\alpha, f(\eta(\alpha), \gamma)) \in A$; if $\beta = \pi(\alpha, \gamma)$, then by the definition

$$F_\beta(\delta) = F(\pi(\beta, \delta)) = f(\eta(\alpha), \gamma)(\delta),$$

i.e. $F_\beta = f(\eta(\alpha), \gamma)$, hence $(\alpha, F_\beta) \in A$.

THEOREM 9. *Assume AD and Collection. Then each $\boldsymbol{\delta}_k^2$ is a regular limit cardinal (inaccessible). Moreover, if $\leqq$ is a $\mathbf{\Delta}_k^2$ prewellordering with length ξ and $A \subseteq \xi$, then $Cod(A; \leqq)$ is $\mathbf{\Delta}_k^2$, hence A is definable (in set theory) from a real parameter.*†

† Another very easy consequence of *Collection* is that $o(^R2)$ is a regular ordinal. Thus with AD and *Collection*, $o(^R2)$ has fairly strong reflection properties, since it is e.g. a regular limit of Mahlo cardinals. (We do not know how to prove with AD and *Collection* that $o(^R2)$ is Mahlo.) Solovay proved first that $o(^R2)$ is regular in $L[R]$ — this follows from our Lemma 11. We doubt that the regularity of $o(^R2)$ can be shown without any choice-like axioms like *Collection.*

Proof is immediate from Theorem 5.

Using the hypotheses of Theorem 9, it is not hard to see that $\boldsymbol{\delta}_1^2$ is a limit of inaccessibles, a limit of regular limits of inaccessibles, etc. However, we can get much better lower bounds for $\boldsymbol{\delta}_1^2$ if we also assume *DC*.

THEOREM 10. *Assume AC, DC, Collection. Let L be the dyadic second order language for an ordering, for each λ let $\mathfrak{A}_\lambda$ be the standard model of L as in §4. Then the first regular κ which reflects for this L, $\{\mathfrak{A}_\lambda\}$ is less than $\boldsymbol{\delta}_1^2$.*†

Proof. Let $\pi: R \times R \to R$ be the canonical recursive homeomorphism, put

$$F \in WF \Leftrightarrow \{(\alpha, \beta): F(\pi(\alpha\ \beta)) = 0\} \text{ is a prewellordering.}$$

Now *DC* implies immediately that $F \in WF$ is an *analytic* relation, i.e. it is defined by a formula of third order number theory without quantifiers over ${}^R\omega$.

If $F \in WF$, let $\leqq_F$ be the prewellordering determined by F, let $\psi_F: Field(\leqq_F) \twoheadrightarrow \lambda_F$ be the canonical surjection onto the length λ_F. Let κ be the least ordinal that reflects with respect to L, $\{\mathfrak{A}_\lambda\}$, let $\phi: \kappa \to {}^R 2$ be the mapping defined in the proof of Theorem 2 which assigns codes to all ordinals less than κ. Put

$$Q(F) \Leftrightarrow F \in WF \;\&\; \lambda_F = \kappa, \tag{7–2}$$

$$R(F, G) \Leftrightarrow F \in WF \;\&\; \forall\alpha\forall\beta[\alpha \in Field(\leqq_F) \Rightarrow [G(\pi(\alpha, \beta)) = 0 \Leftrightarrow \beta \in \phi(\psi_F(\alpha))]]. \tag{7–3}$$

We wish to show that both $Q(F)$ and $R(F, G)$ are Δ_1^2, i.e. they are definable both by Σ_1^2 and Π_1^2 formulas of third order number theory. If we can do this we will have proved the theorem, since then the prewellordering $\leqq_\kappa$ of the proof of Theorem 2 will be given by

$$\begin{aligned} \alpha \leqq_\kappa \beta \Leftrightarrow\ & \exists F \exists G[R(F, G) \;\&\; \exists\gamma\exists\delta[F(\pi(\gamma, \delta)) = 0 \\ & \;\&\; G(\pi(\gamma, \alpha)) = 0 \;\&\; G(\pi(\delta, \beta)) = 0]] \\ \Leftrightarrow\ & \forall F \forall G[Q(F) \;\&\; R(F, G) \Rightarrow \exists\gamma\exists\delta[F(\pi(\gamma, \delta)) = 0 \\ & \;\&\; G(\pi(\gamma, \alpha)) = 0 \;\&\; G(\pi(\delta, \beta)) = 0]]. \end{aligned}$$

We shall outline the computation of $Q(F)$, $R(F, G)$ in a sequence of sublemmas.

† This result was obtained after the Jerusalem meeting. At the same time Solovay also proved independently and by a different method that (with *AD*) there are highly Mahlo cardinals less than $\boldsymbol{\delta}_1^2$.

SUBLEMMA 10.1. *Put*

$$P_1(n,m,F,G,) \Leftrightarrow [n \text{ is the Gödel number of a formula } \theta(\leqq^*, A^*, \alpha_1, \dots, \alpha_m) \text{ analytic in the binary symbol } \leqq^* \text{ and the unary symbol } A^* \text{ and with } m \text{ free variables}] \;\&\; [\theta(\leqq^*, A^*, (\alpha)_1, \dots, (\alpha)_m) \text{ is true when we interpret } \leqq^* \text{ by } \leqq_F \text{ and } A^* \text{ by } \{\beta : G(\beta) = 0\}].$$

Then $P_1(n,m,F,G,\alpha)$ *is a* Δ_1^2 *relation.*

Proof of this sublemma is by the usual analysis of the induction involved in the definition of truth and we shall omit it.

SUBLEMMA 10.2. *Let* $H(F)$ *be the canonical universal set for* $\Sigma_1^1(\leqq_F)$, *put*

$$P_2(F,\beta) \Leftrightarrow \text{for some } B \subseteq \lambda_F,\ H_\beta(F) = Cod(B; \leqq_F).$$

Then $P_2(F,\beta)$ *is analytic.*

Proof. $H(F)$ is analytically definable from F and then

$$P_2(F,\beta) \Leftrightarrow \forall\alpha[\alpha \in H_\beta(F) \Rightarrow \alpha \in Field(\leqq_F)]$$
$$\& \ \forall\alpha\forall\gamma[\alpha \in H_\beta(F) \,\&\, \alpha \leqq_F \gamma \,\&\, \gamma \leqq_F \alpha \Rightarrow \gamma \in H_\beta(F)].$$

SUBLEMMA 10.3. *Put*

$$P_3(n,m,F\ \beta,\alpha) \Leftrightarrow [n \text{ is the Gödel number of a formula } \theta(x_1, \dots, x_m, A) \text{ of the dyadic second order language for one ordering with } m \text{ free individual variables and one free set variable}]$$

$$\& \ F \in WF \ \& \ \forall i[1 \leqq i \leqq m \Rightarrow (\alpha)_i \in Field\ (\leqq_F)] \ \& \ P_2(F,\beta)$$

$$\& \ [\mathfrak{A}_{\lambda_F} \models \theta(\psi_F((\alpha)_m), \dots, \psi_F((\alpha)_m), B), \text{ when } B \text{ is such that } H_\beta(F) = Cod(B; \leqq_F)].$$

Then $P_3(n,m,F,\beta,\alpha)$ *is* Δ_1^2.

Proof is immediate from Sublemmas 10.1, 10.2 and Lemma 8.

SUBLEMMA 10.4. *Put*

$$P_4(F,\beta,\gamma) \Leftrightarrow F \in WF \,\&\, \gamma \in Field(\leqq_F) \,\&\, P_2(F,\beta)$$
$$\& \ [\text{if } B \subseteq \lambda_F \text{ is such that } H_\beta(F) = Cod(B; \leqq_F),$$
$$\text{then } (\psi_F(\gamma),\ B \cap \psi_F(\gamma)) \prec\!\prec (\lambda_F, B)].$$

Then $P_4(F,\beta,\alpha)$ *is* Δ_1^2.

Proof. The last condition in $P_4(F,\beta,\gamma)$ is equivalent to

$$\forall n \forall m \forall \alpha \{[\forall i[1 \leqq i \leqq m \Rightarrow [(\alpha)_i \leqq_F \gamma \ \& \ \neg \gamma \leqq_F (\alpha)_i]] \ \& \ P_3(n,m,F^\gamma,\beta^\gamma,\alpha)] \Rightarrow P_3(n,m,F,\beta,\alpha)\},$$

where F^γ, β^γ are chosen so that $\leqq_{F_\gamma}$ is the restriction of $\leqq_F$ to the points less than γ and if $H_\beta(F) = Cod(B; \leqq_F)$, then

$$H_{\beta^\gamma}(F^\gamma) = Cod(B \cap \psi_F(\gamma); \leqq_{F^\gamma}).$$

SUBLEMMA 10.5. *The condition $Q(F)$ defined by (7–2) is Δ_1^2.*

Proof. Using the preceding sublemmas, put

$$P_5(F) \Leftrightarrow F \in WF \ \& \ \lambda_F \text{ is regular}$$
$$\& \ \forall \beta \forall \gamma [P_2(F,\beta) \ \& \ \gamma \in Field(\leqq_F)$$
$$\Rightarrow \exists \gamma'[\gamma \leqq_F \gamma' \ \& \ P_4(F,\beta,\gamma')]].$$

Notice that the condition "λ_F *is regular*" is Δ_1^2, since it is equivalent to $P_3(n_0,\theta,F,\beta_0,\alpha_0)$ for some fixed n_0,β_0,α_0, so that $P_5(F)$ is Δ_1^2. Now

$$Q(F) \Leftrightarrow P_5(F) \ \& \ \forall \gamma [\gamma \in Field(\leqq_F) \Rightarrow \neg P_5(F^\gamma)].$$

The mapping $\phi: \kappa \to {}^R 2$ was defined by transfinite induction on $\xi < \kappa$, where for each $\xi < \kappa$ there were four cases. In order to prove that $R(F,G)$ is Δ_1^2, we must show that each of the case hypotheses, and then the definition in each case are Δ_1^2 (in terms of the coding of ordinals provided by F). One can write down all the clauses quite easily, by applying the sublemmas above, but it is a tedious mess and we shall avoid committing it to print.

This proof of Theorem 10 can be directly relativized to any given prewellordering in Δ_1^2 so that it yields the following result: *for each $\xi < \boldsymbol{\delta}_1^2$, there is some κ, $\xi \leqq \kappa < \boldsymbol{\delta}_1^2$, such that κ reflects with respect to L, $\{\mathfrak{A}_\lambda\}$.* In particular, $\boldsymbol{\delta}_1^2$ is a limit of ordinals κ which are in $\mathfrak{M}_\kappa$.

Perhaps we should remark that Theorem 9 is easily extended to the classes $\boldsymbol{\Delta}_k^n$, with $n > 2$. It is easy to formulate the appropriate collection property which allows us to prove that $\boldsymbol{\Sigma}_k^n, \boldsymbol{\Pi}_k^n$ are closed under $\exists m$, $\forall m$, $\exists \alpha$, $\forall \alpha$ and then show that it holds in $L[R]$; this implies then that $\boldsymbol{\delta}_k^n = o(\boldsymbol{\Delta}_k^n)$ $(n \geqq 2,\ k \geqq 1)$ is a regular cardinal.

REFERENCES

[1] J. W. ADDISON, *Separation principles in the hierarchies of classical and effective descriptive set theory*, Fund. Math., vol. **46** (1959a), pp. 123–135.

[2] ———, *Some consequences of the axiom of constructibility*, Fund. Math. **46** (1959b), pp. 337–357.

[3] J. W. Addison and Yiannis N. Moschovakis, *Some consequences of the axiom of definable determinateness*, Proc. Nat. Acad. Sc. USA **59** (1968), pp. 708–712.

[4] D. Gale and F. M. Stewart, *Infinite games with perfect information*, Ann. Math. Studies **28** (153), pp. 245–266.

[5] K. Gödel, *The consistency of the axiom of choice and of the generalized continuum hypothesis with the axioms of set theory*, Ann. Math. Studies, No. 3, Princeton University Press, Princeton, N. J. 1940.

[6] S. C. Kleene, *Introduction to metamathematics*, Van Nostrand, Princeton, N. J. 1952.

[7] ———, *Countable functionals*, Constructivity in mathematics, North-Holland, Amsterdam 1955.

[8] D. A. Martin, *The axiom of determinateness and reduction principles in the analytical hierarchy*, Bulletin Amer. Math. Soc. **74** (1968), pp. 687–689.

[9] Y. N. Moschovakis, *Hyperanalytic predicates*, Trans. Amer. Math. Soc. **129** (1967), pp. 249–282.

[10] ———, *Abstract first order computability*. II, Trans. Amer. Math. Soc. **138** (1969), pp. 465–504.

[11] J. Mycielski, *On the axiom of determinateness*, Fund. Math. **53** (1964), pp. 205–224.

[12] H. Rogers Jr., *Theory of recursive functions and effective computability*, McGraw-Hill, New York, 1967.

[13] Y. Suzuki, *A complete classification of the* Δ_2^1 *functions*, Bull. Amer. Math. Soc. **70** (1964), pp. 246–253.

INITIAL SEGMENTS OF THE DEGREES OF UNSOLVABILITY

PART I: A SURVEY

C. E. M. YATES
The University, Manchester, England

The theory of initial segments is not combinatorially the most interesting part of the theory of degrees of unsolvability. It has, however, at present the most interesting connections with other branches of Mathematical Logic and in particular with Set Theory. On the one hand, this part of the theory of degrees of unsolvability has inspired a sequence of results in Set Theory on non-constructible sets of integers.† The techniques had first to be appropriately generalised to the less absolute notion of degree of constructibility and this resulted in 'perfect-set forcing', the origins of which will be fairly clear to anyone who is familiar with the construction of minimal degrees of unsolvability. On the other hand, a purely set-theoretic approach is useful in handling some of the problems in this area; indeed a number of them are almost certainly independent of the usual axioms for Set Theory. Certainly, it is not always necessary to have the same sensitive feel for computability which is so essential with some of the more delicate problems in the theory of recursively enumerable degrees.

In view of the purpose of this Colloquium it seems most appropriate here, in this first part of thc paper, to survey the general theory of initial segments of the degrees of unsolvability as it exists at the present time. The subsequent parts of this paper, which will be published elsewhere, are devoted to a number of results on initial segments which *do* involve much of the delicacy mentioned above. These results consist roughly of showing that theorems on initial segments, which have been proved previously, can be made more constructive in some way. The first result of this type, namely that there is a minimal degree below $\mathbf{0}^{(1)}$ (the highest recursively enumerable degree), was obtained some time ago by Sacks [9]; the existence of a minimal degree had been previously obtained by Spector [15], thereby initiating the theory of initial segments. We shall present some much stronger and more difficult theorems of this type. For example, we prove in Part II that every nonzero recursively enumerable degree has a minimal predecessor.

† One of the deepest results of this type can be found in Professor Jensen's contribution to this volume.

This turns out to be in a sense as strong as is possible, because the main result of Part III is that there is a degree below $\mathbf{0}^{(1)}$ which has no minimal predecessor. These theorems answer two questions in Sacks' monograph [10, §12, Q. 1 and Q. 3]. The principal result of Part IV is that any distributive lattice (with least element) of degree $\leqq \mathbf{0}^{(1)}$ can be embedded as an initial segment of the degrees below $\mathbf{0}^{(1)}$; this is the constructive counterpart of the Hugill-Lachlan embedding-theorem for the degrees [4], and it yields the undecidability of the elementary theory of the degrees below $\mathbf{0}^{(1)}$.

We shall separately prove each theorem before proving its more constructive counterpart. This should help to clarify the more complicated proof of the latter which inevitably involves some sort of priority argument. In any case there is some need for an exposition of this area of the subject, since so much has been done since the publication of Sacks' monograph. The following summary of the first four parts of the paper should help describe the ground that will be covered.

Part I: A survey

§1. Introduction to $\boldsymbol{D}$.
§2. Finite initial segments of $\boldsymbol{D}$.
§3. Countable initial segments of $\boldsymbol{D}$.
§4. Initial segments of $\boldsymbol{D}$ ($\leqq 0^{(1)}$).
§5. Initial segments of $\boldsymbol{D}$ ($\leqq \boldsymbol{a}$) for arbitrary recursively enumerable $\boldsymbol{a}$.
§6. Uncountable initial segments of $\boldsymbol{D}$.
§7. Some general problems of isomorphism and indiscernibility.

Part II: Minimal degrees.

§8. Partial recursive trees.
§9. A minimal degree below $\mathbf{0}^{(2)}$.
§10. A minimal degree below $\mathbf{0}^{(1)}$.
§11. A minimal degree below any nonzero recursively enumerable degree.

Part III: Finite Boolean algebras and degrees with no minimal predecessors.

§12. Partial recursive trees again.
§13. Finite Boolean algebras and $\boldsymbol{D}$ ($\leqq \mathbf{0}^{(2)}$).
§14. Finite Boolean algebras and $\boldsymbol{D}$ ($\leqq \mathbf{0}^{(1)}$).
§15. A nonzero degree with no minimal predecessor.
§16. A nonzero degree below $\mathbf{0}^{(1)}$ with no minimal predecessor.

Part IV: Distributive lattices.

§17. Finite distributive lattices and $\boldsymbol{D}$ ($\leqq \mathbf{0}^{(2)}$).
§18. Finite distributive lattices and $\boldsymbol{D}$ ($\leqq \mathbf{0}^{(1)}$).

We plan to devote Parts V and VI to non-distributive and uncountable initial segments respectively.

§1. Introduction to D

This introduction is intended to describe the modern approach to relative recursiveness† together with a few relevant facts about the degrees of unsolvability; in particular, a the end of this section we shall sketch a proof of the existence of minimal degrees. This treatment is by no means intended to be exhaustive and so familiarity is assured with the basic concepts of Recursion Theory; Rogers' book [8] is an excellent reference for such material, and Sacks' monograph [10] should be consulted for any further information on the degrees of unsolvability. We shall depart from certain standard notation which we do not like; for example, we shall use $f_0, f_1, \cdots$, to denote some standard recursive enumeration of the partial recursive functions from N into $\{0,1\}$, where N is the set of natural numbers. We recall that a partial recursive function is recursive if its domain is N.

Following Shoenfield [13] we formulate the notion of relative recursiveness in terms of strings: a *string* is just a finite sequence of zeros and ones. We let S denote the set of all strings and in general we shall use lower-case Greek letters to denote strings. Whenever it is necessary we shall use $\emptyset$ to denote the null string; this should cause no confusion. The number of elements of a string σ will be called the *length* of σ and is denoted by $\mathrm{lh}(\sigma)$. The $n+1$-th element of σ is denoted by $\sigma(n)$ and the initial segment of σ which has length $n+1$ is denoted by $\sigma[n]$. We let $\sigma * \tau$ be the string obtained by adding τ to the right hand end of σ; in particular, $\sigma * 0$ and $\sigma * 1$ are obtained by adding a single zero and one respectively, and $\sigma * \emptyset = \sigma$. If there is any risk of ambiguity, the string consisting of the number n alone will be denoted by 'n'. We shall write $\sigma \subset \sigma'$ to mean that σ' is a proper extension of σ, in other words $\sigma' = \sigma * \tau$ for some $\tau \neq \emptyset$. If $\sigma(n) \neq \sigma'(n)$ for some n then we say that σ and σ' are incompatible, written $\sigma | \tau$, and we let $n(\sigma\ \sigma')$ be the least n such that $\sigma(n) \neq \sigma'(n)$. Since we can identify a set $X \subseteq N$ with its characteristic function and this in turn can be regarded as an infinite sequence of zeros and ones, we also write $\sigma \subset X$ to mean that $\sigma = X[n]$ for some n, where $X[n]$ is the initial segment of the characteristic function of X which has length $n+1$.

There are a number of obvious recursive enumerations of S; the most

† This elegant approach is due essentially the Shoenfield [13].

natural is perhaps the lexicographic enumeration although the prime-power enumeration is frequently used. Anyway, we can discuss recursion-theory on $\boldsymbol{S}$ just as easily as we can on N, and the particular representation chosen in order to lift results directly from the standard theory is irrelevant. It is sufficient to say that we deal with questions of quantification (for example) over $\boldsymbol{S}$ exactly as over $\boldsymbol{N}$. In order to do this it is only necessary to postulate (or prove through representation) the intuitively obvious fact that all the operations and relations concerning strings which we defined above are recursive.

As we have mentioned, the main reason for introducing strings is that they play a significant role in questions of relative recursiveness between subsets of N. This is because in computing the value $X(n)$ from Y, where X and $Y \subseteq N$, it is only necessary to use the information contained in some string $\sigma_n \subset Y$ dependent on n. The definition of relative recursiveness which follows is designed to emphasise this continuity of partial recursive functionals. We begin with a useful subsidiary definition.

DEFINITION 1.1. We say that F *is an* $\boldsymbol{S}$*-map* if its domain and range are subsets of $\boldsymbol{S}$. F is *order-preserving* if $F(\sigma) \subseteq F(\tau)$ for all σ, τ in the domain of F such that $\sigma \subseteq \tau$.

Let F be an order-preserving $\boldsymbol{S}$-map. We define

$$F^*(X) = \bigcup_{\sigma \subset X} F(\sigma),$$

where $X \subseteq \boldsymbol{N}$, with the understanding that $F^*(X)$ is only defined if the R.H.S. is infinite. Let $F_0, F_1, \cdots,$ be a recursive enumeration of all partial recursive $\boldsymbol{S}$-maps. Then $F_0^*, F_1^*, \cdots,$ is essentially an enumeration of all partial recursively functionals. Now we can define relative recursiveness.

DEFINITION 1.2. We say that X is *recursive in* Y, where $X, Y \subseteq \boldsymbol{N}$, if there is an order-preserving partial recursive $\boldsymbol{S}$-map F such that

$$X = F^*(Y).$$

We shall occasionally write $X \leqq_T Y$ to mean that X is recursive in Y.

Intuitively, in order to compute X from Y, we simply compute $F(\sigma)$ for $\sigma \subset Y$ looking for σ which produce increasingly large strings: the limit of these strings is X.

Let $X \equiv_T Y$ mean that $X \leqq_T Y$ and $Y \leqq_T X$; then $\equiv_T$ is an equivalence relation over 2^N because $\leqq_T$ is easily seen to be transitive. The corresponding equivalence classes constitute the set $\boldsymbol{D}$ of degrees of unsolvability. Clearly, $\leqq_T$ induces a partial ordering $\leqq$ of $\boldsymbol{D}$.†

† We use $\boldsymbol{D}(\leqq \boldsymbol{a})$, $\boldsymbol{D}(\geqq \boldsymbol{a})$, $\boldsymbol{D}(< \boldsymbol{a})$ and $\boldsymbol{D}(> \boldsymbol{a})$ as abbreviations for the degrees of unsolvability $\leqq \boldsymbol{a}$, $\geqq \boldsymbol{a}$, $< \boldsymbol{a}$ and $> \boldsymbol{a}$, respectively, for any $\boldsymbol{a} \leqq \boldsymbol{D}$.

We now enumerate some fundamental facts about $\boldsymbol{D}$ which are relevant in the sequel. First, we recall that a degree $\boldsymbol{b}$ is recursively enumerable in a degree $\boldsymbol{a}$ if some set of degree $\boldsymbol{b}$ is recursively enumerable (r.e.) in some set (and hence every set) of degree $\boldsymbol{a}$. Also, we say that a partially-ordered (p.o.) set possesses the countable predecessors property (c.p.p.) if each element has countably many predecessors, i.e. elements $\leqq$ to it.

(i) $\boldsymbol{D}$ (and hence any subset of $\boldsymbol{D}$) under $\leqq$ possesses the c.p.p.

(ii) $\boldsymbol{D}$ has a least element $\boldsymbol{0}$, the degree of the recursive sets.

(iii) $\boldsymbol{D}$ is an upper semi-lattice: the join or l.u.b. of each two elements of $\boldsymbol{D}$ exists and can be characterised in a very simple manner.

(iv) Let a degree $\boldsymbol{a}$ be called minimal if $\boldsymbol{a}$ is nonzero but $\boldsymbol{0}$ is the only degree $< \boldsymbol{a}$. Then by a theorem of Spector (extended slightly by Lacombe) there are $2^{\boldsymbol{N}}$ minimal degrees. (As we have mentioned we shall sketch a proof of this at the end of this section.) Notice that any two minimal degrees have g.l.b. $\boldsymbol{0}$. (But g.l.b.'s do not exist in general, i.e. $\boldsymbol{D}$ is *not* a lattice; there are now several proofs of this.)

(v) Any countable ascending sequence of degrees has $2^{\boldsymbol{N}}$ minimal upper bounds; this is an extension of Spector's theorem due to Sacks [10].

(vi) for each $\boldsymbol{a} \in \boldsymbol{D}$ there is a degree $\boldsymbol{a}^{(1)}$ (the *jump* of $\boldsymbol{a}$) such that $\boldsymbol{a} < \boldsymbol{a}^{(1)}$; $\boldsymbol{a}^{(1)}$ is r.e. in $\boldsymbol{a}$ and every degree which is r.e. in $\boldsymbol{a}$ is $\leqq \boldsymbol{a}^{(1)}$. The converse of the last statement is far from true; for example, there are no minimal r.e. degrees but there is a minimal degree $\leqq \boldsymbol{0}^{(1)}$. Subsequent parts of this paper will provide many additional counterexamples. We shall write $\boldsymbol{a}^{(n)}$ to denote the result of iterating the jump operation n times. This yields an increasing sequence of degrees beginning with $\boldsymbol{a}$ which can in fact be continued through all the countable ordinals.

(vii) Since each degree contains only countably many sets, the Axiom of Choice immediately implies that there are exactly $2^{\boldsymbol{N}}$ degrees. In fact, an injection from $2^{\boldsymbol{N}}$ to $\boldsymbol{D}$ can be defined without the Axiom of Choice. On the other hand, it is consistent with Set Theory (excluding the Axiom of Choice) that there is no injection from $\boldsymbol{D}$ to $2^{\boldsymbol{N}}$, i.e. there are more degrees than reals! This typical paradox is most quickly deduced from Solovay's work on translation-invariant extensions of Lebesgue-measure, from which it follows that there is a 2-valued countably additive measure on $\boldsymbol{D}$: there can be no such measure on $2^{\boldsymbol{N}}$.† (An elegant exposition of Solovay's work can be found in Sacks [11].)

The observations above should give some general feeling for the structure $\boldsymbol{D}$.

Results in the theory of initial segments consist almost entirely of showing that certain p.o. sets can be embedded as initial segments of $\boldsymbol{D}$ (or some

† A more direct proof was verbally communicated to the writer by Professor Solovay at the Colloquium.

subset of $\boldsymbol{D}$). To make the discussion quite precise from here on we need the following definitions.

DEFINITION 1.3. Let $(D, \leqq)$ be a p.o. set. A p.o. set $(I, \leqq')$ is said to be an *initial segment* of $(D, \leqq)$ if:

(i) $I \subseteq D$ and $\leqq'$ is the restriction of $\leqq$ to I;

(ii) $x \leqq y$ and $y \in I$ implies that $x \in I$.

An initial segment is said to be *topped* if it has a largest element; otherwise it is called *topless*. It is said to be *closed* if for all $x, y \in I$ there is a $z \in I$ such that $x \leqq z$ and $y \leqq z$. Every topped segment is obviously closed.

In most situations, $\leqq$ is the partial ordering $\leqq$ of $\boldsymbol{D}$ and so will be omitted. Notice that every closed initial segment of $\boldsymbol{D}$ is an upper semilattice, and every topped initial segment is countable (because $\boldsymbol{D}$ has the c.p.p.). Also, every initial segment of $\boldsymbol{D}$ is of course *bottomed*, i.e. possesses a least element; hence we shall not be interested in bottomless structures.

DEFINITION 1.4. A p.o. set $(\rho, \leqq')$ is *embeddable in* a p.o. set $(D, \leqq)$, written $(P, \leqq') \Rightarrow (D, \leqq)$, if there is an order-isomorphism between $(\rho, \leqq')$ and $(Q, \leqq)$ for some $Q \subseteq D$. $(P, \leqq')$ is *embeddable as an initial segment of* $(D, \leqq)$, written $(P, \leqq') \overset{*}{\Rightarrow} (D, \leqq)$, if there is an order-isomorphism between $(P, \leqq')$ and $(I, \leqq)$ for some initial segment I of D.

We shall frequently abuse the terminology of Definition 1.4 by omitting the order relations as long as it is clear what they are. In the present paper, we are not particularly interested in $\Rightarrow$, but it is worth recalling that Sacks [10, §3] has proved that if $(P, \leqq')$ is a p.o. set of cardinality $\leqq \aleph_1$ having the c.c.p. then $(P, \leqq') \Rightarrow \boldsymbol{D}$. It is still an open question whether $\aleph_1$ can be replaced here by $2^{\aleph_0}$ (cf. conjecture C.4. of §12 of [10]), but it seems likely that this is independent of the usual axioms for Set Theory; it is of course consistent since it is immediately implied by the Continuum Hypothesis.

We conclude this introduction by sketching a proof of the existence of minimal degrees (mentioned in (iv) above), since this can be done very briefly and is the cornerstone of the theory of initial segments. The following definitions are adequate for this proof although we shall need more general definitions for the subsequent parts of the paper.

DEFINITION 1.5. An $\boldsymbol{S}$-map F is an $\boldsymbol{S}$-*isomorphism* if it is an injection and

$$\sigma \subseteq \tau \quad \leftrightarrow \quad F(\sigma) \subseteq F(\tau)$$

for all $\sigma, \tau \in \boldsymbol{S}$. A *tree* is an $\boldsymbol{S}$-isomorphism with domain $\boldsymbol{S}$. T' is a *subtree* of a tree T if $\mathrm{im}(T') \subseteq \mathrm{im}(T)$.

If F is an $\boldsymbol{S}$-isomorphism then F^* is an injection. If F is also partial recursive then both F^* and $(F^*)^{-1}$ are partial recursive functionals and so X and $F^*(X)$ have the same degree whenever $X \in \text{dom}(F^*)$.

DEFINITION 1.6. A pair of strings (σ_0, σ_1) *splits a string* σ *for* e if $\sigma_0 \supseteq \sigma$, $\sigma_1 \supseteq \sigma$ and $\boldsymbol{F}_e(\sigma_0)$, $\boldsymbol{F}_e(\sigma_1)$ are both defined and incompatible. (Since $\boldsymbol{F}_e$ is an order-preserving $\boldsymbol{S}$-map, for each e, it follows that σ_0, σ_1 are incompatible.) A tree T is a *splitting tree for* e if $T(\tau * 0)$, $T(\tau * 1)$ split $T(\tau)$ for e, for all τ.

It is now easy to prove the following two lemmas, which represent the core of Shoenfield's analysis of the construction of minimal degrees [13].

LEMMA A. *Let T be a recursive tree. If $Y \in \text{im}(T^*) \cap \text{dom}(\boldsymbol{F}_e^*)$ and β is a substring of Y in* im(T) *such that no two strings in* im(T) *split β for e, then $\boldsymbol{F}_e^*(Y)$ is recursive.*

LEMMA B. *Let T be a recursive splitting tree for e. If $Y \in \text{im}(T^*)$ then $Y \in \text{dom}(\boldsymbol{F}_e^*)$ and Y has the same degree of unsolvability as $\boldsymbol{F}_e^*(Y)$.*

The proof of the latter lemma consists simply of observing that $\boldsymbol{F}_e$ restricted to im(T) is a partial recursive $\boldsymbol{S}$-isomorphism.

Now we are in a position to sketch a proof of the existence of minimal degrees, in fact of the following rather stronger theorem.

THEOREM 1.7 (Spector-Lacombe). *There are $2^{\boldsymbol{N}}$ minimal degrees.*

PROOF. Using Lemmas A and B we can define an array of recursive trees $\{T_\sigma : \sigma \in \boldsymbol{S}\}$ such that $T_{\sigma * 0}$, $T_{\sigma * 1}$ are disjoint subtrees of T_σ for all $\sigma \in \boldsymbol{S}$; moreover, either $T_{\sigma * i}$ is a splitting tree for lh(σ) or no pair of extensions of $T_{\sigma * i}\ (\phi)$ in im(T_σ) split $T_{\sigma * i}(\phi)$ for lh(σ). It is easy to prove that if $\sigma_0 \subset \sigma_1 \subset \cdots$ is any ascending chain of strings, such that $\sigma_{n+1} = \sigma_n * i$ for some i dependent on n, then the unique set in the intersection of the contracting chain

$$T_{\sigma_0} \supset T_{\sigma_1} \supset \cdots$$

is either recursive or of minimal degree. There are in fact $2^{\boldsymbol{N}}$ possibilities for this sequence, and it easily follows that there are $2^{\boldsymbol{N}}$ minimal degrees. ∎

It is worthwhile listing some of the corollaries of this theorem, since it yields a greater number of interesting corollaries than do most theorems on degrees. First, we observe that any two minimal degrees are incomparable, and so the minimal degrees form a continuum of mutually incomparable degrees.

COROLLARY 1.8. *If $\boldsymbol{a}_0, \boldsymbol{a}_1, \ldots,$ is a sequence of nonzero degrees then there is a continuum of mutually incomparable degrees, each of which is incomparable with every degree $\boldsymbol{a}_n$.*

In particular, of course, given any degree $\boldsymbol{a} > \mathbf{0}$ there is a minimal degree $\boldsymbol{b}$ which is incomparable with $\boldsymbol{a}$. Another immediately obvious corollary is:

COROLLARY 1.9. *If* $\boldsymbol{a}$ *is any nonzero degree then there are* 2^N *degrees* $\boldsymbol{b}$ *such that the g.l.b. of* $\boldsymbol{a}$ *and* $\boldsymbol{b}$ *exists and is* $\mathbf{0}$.

Every degree which is r.e. in some lower degree is known by a result of Sacks [10, §5] to be the join of two incomparable degrees. On the other hand, it follows from Theorem 1.7 that:

COROLLARY 1.10. *There are* 2^N *degrees* $\boldsymbol{b}$ *such that* $\boldsymbol{b}$ *is not the join of two incomparable degrees.*

There are many generalisations of Theorem 1.7, or at least of the construction. The most obvious of these is that if $\boldsymbol{d}$ is any degree then there are 2^N degrees $\boldsymbol{b}$ such that $\boldsymbol{d}<\boldsymbol{b}$ but $\boldsymbol{d}<\boldsymbol{c}<\boldsymbol{b}$ for *no* degree $\boldsymbol{c}$. (It is worth noting that this can *not* be strengthened to read "but $\boldsymbol{c} < \boldsymbol{b}$ implies that $\boldsymbol{c} \leqq \boldsymbol{d}$"; for, if $\boldsymbol{d} = \mathbf{0}^{(1)}$ then $\boldsymbol{b} = \boldsymbol{x}^{(1)}$ for some degree $\boldsymbol{x} < \boldsymbol{b}$ and so $\boldsymbol{b}$ is the join of two incomparable degrees which cannot both be $\leqq \boldsymbol{d}$.) A yet stronger theorem is the pleasing result of Sacks mentioned in (v) above: every ascending sequence of degrees has 2^N minimal upper bounds.

Another fruitful line of generalisation is the following. First, observe that the existence of a minimal degree is equivalent to the embeddability of the two-element Boolean algebra as an initial segment of the degrees. It is natural to ask exactly which p.o. sets can be embedded in this way: our purpose in the next four sections is to survey what has been done in this direction.

§2. Finite initial segments of $\boldsymbol{D}$.

Any finite initial segment of $\boldsymbol{D}$ is a subsegment of an initial segment which is a finite bottomed upper semi-lattice. It is natural to ask whether this provides a necessary and sufficient condition for a finite p.o. set to be $\overset{*}{\Rightarrow} \boldsymbol{D}$. In other words, is it true that if $(P, \leqq')$ is a finite p.o. set then $(P, \leqq') \Rightarrow \boldsymbol{D}$ if and only if there is a finite bottomed upper semi-lattice $(L, \leqq)$ such that $(P, \leqq') \overset{*}{\Rightarrow} (L, \leqq)$? Sacks has conjectured [10, §12, c. 6] a positive answer to this question. Since any finite bottomed upper semi-lattice is easily seen to be a lattice, an equivalent formulation of this conjecture is:

Sacks' conjecture: If $(L, \leqq)$ is a finite lattice then $(L, \leqq) \overset{*}{\Rightarrow} \boldsymbol{D}$.

This is the principal open question in the theory of initial segments. A negative answer immediately implies a negative answer to the more general questions which we describe in the next section. Conversely, a positive answer to this question is a first step towards positive answers to a number of questions. Although the problem is open, some definite progress has been made. The first main result that has been obtained is the following.

THEOREM 2.1 (Spector-Titgemeyer-Sacks-Shoenfield-Lachlan). *Every finite distributive lattice is* $\overset{*}{\Rightarrow} \boldsymbol{D}$.

The reason for the extensive distribution of credit is that the theorem was obtained through a number of natural stages over a period of time. As we have already mentioned, it all began with Spector's proof around 1956 that there is a minimal degree of unsolvability, i.e. that the two-element Boolean algebra is $\overset{*}{\Rightarrow} \boldsymbol{D}$. Around 1962 Titgemeyer [17] extended Spector's theorem by proving that every finite chain is $\overset{*}{\Rightarrow} \boldsymbol{D}$; this needed a (by no means obvious) sophistication of Spector's technique, which has been the main contribution to all subsequent work on initial segments. Sacks [10, §11] and Shoenfield (unpublished) abstracted the method further to deal with any finite Boolean algebra. Finally, Lachlan obtained the theorem above by further slight modification; there is some complication in moving from Boolean algebras to arbitrary distributive lattices, but intuitively this is no more than dealing with the omission of certain relative complements and no new ideas are involved. Lachlan in fact obtained (in conjunction with his student Hugill) a stronger theorem which will be discussed in §3 and proved in §19.

Some good progress on non-distributive lattices has been made recently by Shoenfield and especially by Lerman.† They have shown that certain special lattices of this type are $\overset{*}{\Rightarrow} \boldsymbol{D}$. These lattices are best described by means of the following diagrams:

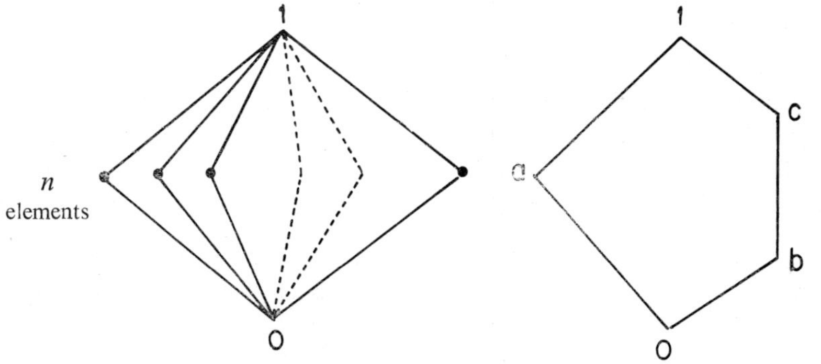

(a) b)

Fig. 1

† Some further generalisations have more recently been obtained by S. K. Thomason (*Sublattices and initial segments of the degrees of unsolvability*, to appear.)

Fig. 1(a) describes a lattice which contains exactly n mutually incomparable elements in addition to its top and bottom elements; this we shall call the nth Chinese lantern.† The structure illustrated in Fig. 1(b) is quite clear: it has three elements a, b, c in addition to its top and bottom elements and these are such that $b \leqq c$ while a is incomparable with both b and c. Following Lerman [5] we shall refer to it as the Pentagon. Shoenfield proved (in an unpublished manuscript) that the Third Chinese Lantern is $\overset{*}{\Rightarrow} \boldsymbol{D}$. More recently, however, Lerman [5] has extended Shoenfield's method to prove the following interesting and much more general theorem:

THEOREM 2.2 (Lerman). *If $n = p^k + 1$ for some prime number p then the n-th Chinese Lantern is $\overset{*}{\Rightarrow} \boldsymbol{D}$.*

The proof uses the fact that if $n = p^k + 1$ for some prime number p then there is a finite projective plane with a line which has exactly n points on it; Shoenfield had essentially used this fact in the special case $n = 3$. The existence of finite projective planes of the required type for numbers which are not the successors of prime powers remains an open question. It is known, however, that there is no such plane in the case $n = 7$, and so Lerman suggests that the Seventh Chinese Lantern may be an example of a finite lattice which is *not* $\overset{*}{\Rightarrow} \boldsymbol{D}$. Certainly this is the first piece of evidence that Sacks' conjecture may be false. Another result obtained by Lerman is the following.

THEOREM 2.3 (Lerman). *The Pentagon is $\overset{*}{\Rightarrow} \boldsymbol{D}$.*

A lattice is non-distributive if and only if it contains either the Third Chinese Lantern or the Pentagon as a sublattice. This is not enough, however, to generalise Lerman's results to arbitrary finite non-distributive lattices.

In §13 of Part III of the present work we shall prove that every finite Boolean algebra is $\overset{*}{\Rightarrow} \boldsymbol{D}$, since it is the most natural basic result to begin with; in §14 we strengthen this result by replacing $\boldsymbol{D}$ by $\boldsymbol{D}\ (\leqq \mathbf{0}^{(1)})$—the degrees $\leqq \mathbf{0}^{(1)}$. The generalisations to distributive lattices are left to a more appropriate place in Part IV. We do not at present know whether Lerman's results extend to $\boldsymbol{D}\ (\leqq \mathbf{0}^{(1)})$, but it would be very surprising if this were not so.

§3. Countable initial segments of $\boldsymbol{D}$.

Any countable initial segment of $\boldsymbol{D}$ is a subsegment of a closed initial segment, which is of course a countable bottomed upper semi-lattice. Again it is natural to ask whether this provides a necessary and sufficient condition for a countable p.o. set to be $\overset{*}{\Rightarrow} \boldsymbol{D}$. In other words, is it true that if $(P, \leqq')$ is a countable p.o. set then $(p, \leqq') \overset{*}{\Rightarrow} \boldsymbol{D}$ if and only if there is a countable bottomed upper semi-lattice $(L, \leqq)$ such that $(p, \leqq') \overset{*}{\Rightarrow} (L, \leqq)$? Lachlan has conjectured a positive answer to this question, thereby ex-

† This picturesque term is due to Professor Robin Gandy.

tending Sacks' conjecture which we discussed in §2. The conjecture is best phrased as follows.

LACHLAN'S CONJECTURE: *If* $(L, \leqq)$ *is a countable bottomed upper semi-lattice then* $(L, \leqq) \overset{*}{\Rightarrow} \boldsymbol{D}$.

It is clear that we could restrict our attention to topped upper semi-lattices, and in proving theorems about countable initial segments it is convenient to do this. There is no need, however, to do so in the present section. In contrast to the finite case, the following weaker conjecture may be distinguished.

SUBSIDIARY CONJECTURE: *If* $(L, \leqq)$ *is a bottomed countable lattice then* $(L, \leqq)$ *is* $\overset{*}{\Rightarrow} \boldsymbol{D}$.

This is of course again an extension of Sacks' conjecture and so remains an open question. It is easy to see that it is weaker than Lachlan's conjecture: for example, the countable upper semi-lattice in Fig. 2 is not a

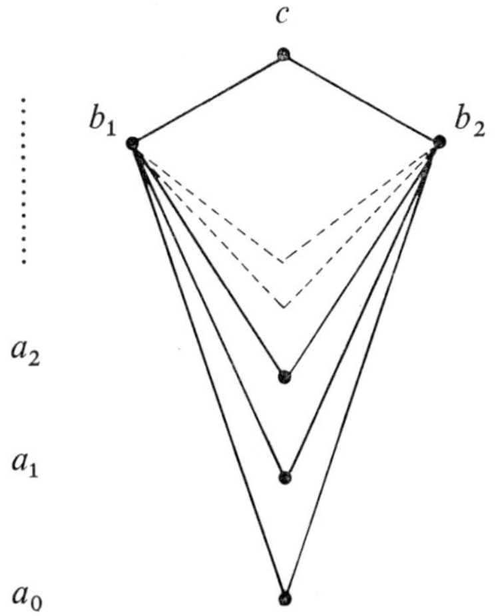

Fig. 2

lattice. Its elements consist of an infinite ascending sequence $a_0, a_1, \cdots$, each element of which is less than both elements of a pair b_1, b_2, which are incomparable and such that $b_1 \cup b_2 = c$. The g.l.b. of b_1 and b_2 clearly does not exist. It is natural to ask whether this simple structure is $\overset{*}{\Rightarrow} \boldsymbol{D}$; at present we do not know the answer but it may not be too hard to find it.

Once again some definite progress has been made. The best result so far obtained is the next theorem, which has a number of interesting corollaries.

THEOREM 3.1 (Hugill-Lachlan). *If* $(L, \leqq)$ *is a bottomed countable distributive lattice then* $(L, \leqq) \overset{*}{\Rightarrow} \boldsymbol{D}$.

The first essential step was made by Hugill [3] who proved the first corollary below. Lachlan then showed how to extend Hugill's method to obtain the general theorem above, and this is contained in their joint paper [4].

COROLLARY 3.2. *If* $(L, \leqq)$ *is a bottomed countable chain then* $(L, \leqq)$ *is* $\overset{*}{\Rightarrow} \boldsymbol{D}$.

An interesting special case of Corollary 3.2 is

COROLLARY 3.3. *If* α *is any countable ordinal (with the usual ordering) then* $\alpha \overset{*}{\Rightarrow} \boldsymbol{D}$.

The next corollary provides yet another particular case of the main theorem which is of special interest.

COROLLARY 3.4. *If* $(B, \leqq)$ *is a countable Boolean algebra then* $(B, \leqq)$ *is* $\overset{*}{\Rightarrow} \boldsymbol{D}$.

Now we turn to some less immediate corollaries. The first of these was obtained first by the present writer using a different method; this was announced in [18]. In fact we obtained the stronger result in which $\boldsymbol{D}$ is replaced by $\boldsymbol{D}\ (\leqq \boldsymbol{0}^{(1)})$.

COROLLARY 3.5. *There is a nonzero degree which has no minimal predecessor.*

Proof. Consider the bottomed countable chain of type $(1, -\omega)$ illustrated in Fig. 3.

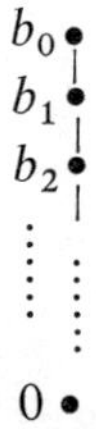

Fig. 3.

It consists of a least element 0 and an infinite descending sequence $b_0, b_1, \ldots$. It follows from Corollary 3.2 that this chain is $\overset{*}{\Rightarrow} \boldsymbol{D}$. Clearly, none of the resulting degrees $\boldsymbol{b}_0, \boldsymbol{b}_1, \ldots$, has a minimal predecessor since no $\boldsymbol{b}_i$ is minimal. ■

This answers a question of Sacks [10, §12, Q. 3]. In addition to the proofs obtained by Hugill and the writer, yet another proof was produced by Martin [6]. He proved an elegant general theorem: if $\mathscr{A}$ is a meager set of degrees then the set of degrees with predecessors in $\mathscr{A}$ is also meager, i.e. the set $\{\boldsymbol{b}:(\exists \boldsymbol{a})(\boldsymbol{a} \leqq \boldsymbol{b}\ \&\ \boldsymbol{a} \in \mathscr{A})\}$ is meager. Since the minimal degrees are well-known to constitute a meager set, it follows that the set of degrees with no minimal predecessors is co-meager and so certainly nonempty. Unfortunately, there does not seem to be any constructive counterpart to Martin's result which would yield our more constructive result; this is one of the shortcomings of what are usually called 'category-methods'. The Hugill-Lachlan result, on the other hand, *can* be constructivised and this is the main objective of Part IV of the present paper.

Before stating the next corollary we recall from §1 that every ascending sequence of degrees has many minimal upper bonds. A natural question to ask is: if $\boldsymbol{a}_0 < \boldsymbol{a}_1 < \cdots < \boldsymbol{b}$, does the ascending sequence $\boldsymbol{a}_0 < \boldsymbol{a}_1 < \ldots$ have a minimal upper bound $\leqq \boldsymbol{b}$?† The following corollary shows that the answer is 'no' in general.

COROLLARY 3.6. *There is an ascending sequence of degrees* $\boldsymbol{a}_0 < \boldsymbol{a}_1 < \ldots$ *which has an upper bound* $\boldsymbol{b}$ *with no minimal upper bound* $\leqq \boldsymbol{b}$.

Proof. Consider the bottomed countable chain of type $(\omega, -\omega)$ illustrated in Fig. 4

Fig. 4.

It consists of an infinite descending sequence $b > b_1 > b_2 > \cdots$ above an infinite ascending sequence $0 < a_1 < a_2 < \ldots$. It follows from Corollary 3.2 that this chain is $\stackrel{*}{\Rightarrow} \boldsymbol{D}$. Clearly, $\boldsymbol{a}_1 < \boldsymbol{a}_2 < \ldots$ has no minimal upper bound $\leqq' \boldsymbol{b}$. ■

We do not know whether this corollary can be improved to replace $\boldsymbol{b}$ by some fixed degree such as $\boldsymbol{0}^{(1)}$.

Perhaps the most interesting corollary of Theorem 3.1 is the next and last one; it was pointed out by Hugill and Lachlan in [4]. Let **EDL** and **EDU** be the (axiomatisable) theory of distributive lattices and the (semantically defined) theory of degrees of unsolvability respectively. Both theories are to be thought of as formulated in $\mathscr{L}_{\leqq}$: elementary logic with equality

† This question was proposed by Dr. Joan Moschovakis.

and a single additional binary relation $\leqq$. **EDU** is the set of all sentences of $\mathscr{L}_{\leqq}$ which are valid in the structure $(\boldsymbol{D}, \leqq)$.

COROLLARY 3.7. **EDL** *is recursive in* **EDU**, *and hence* **EDU** *is not axiomatisable.*

Proof. Let Φ be any sentence of $\mathscr{L}_{\leqq}$. Then

$$\neg(\mathbf{EDL} \vdash \Phi) \leftrightarrow (\mathbf{EDL} \cup \{\neg \Phi\} \text{ has a countable model})$$

by the Completeness Theorem for elementary logic. Now, any distributive lattice can be completed to form a topped and bottomed distributive lattice simply by the adjunction of two elements and the obvious relations between them and the original elements. Hence, by Theorem 3.1 it follows that

$$\neg(\mathbf{EDL} \vdash \Phi) \leftrightarrow (\exists \boldsymbol{a})(\boldsymbol{D}(>\mathbf{0}) \cap \boldsymbol{D}(<\boldsymbol{a}) \models \neg \Phi);$$

we are forced to choose $\boldsymbol{D}(>\mathbf{0}) \cap \boldsymbol{D}(<\boldsymbol{a})$ here because $\neg\Phi$ may imply either "$(\forall x)(\exists y)(x<y)$" or "$(\forall x)(\exists y)(y<x)$" or both. Next, it can be seen that

$$\neg(\mathbf{EDL} \vdash \Phi) \leftrightarrow \mathbf{EDU} \vdash (\exists x)\Phi'(x),$$

where $\Phi'(x)$ is the formula obtained by restricting all the quantified variables in $\neg\Phi$ to be $\leqq x$. Since Φ' can be computed effectively from Φ, this shows that **EDL** is recursive in **EDU**.

Since Grzegorczyk [2] has shown **EDL** to be undecidable, it follows that **EDU** is undecidable. But **EDU** is a complete theory and so it is not even axiomatisable. ∎

This completes the consequences of the Hugill-Lachlan theorem which seem at present to be of interest. We shall prove the theorem in Part IV and show further how to obtain an analogous result for $\boldsymbol{D}(\leqq \mathbf{0}^{(1)})$. Further discussion of this can be found in the next section.

The results of Lerman that were discussed in §2 can be used to show that a number of odd countable nondistributive lattices are $\overset{*}{\Rightarrow} \boldsymbol{D}$. Until Sacks' conjecture is settled one way or the other, however, the Hugill-Lachlan theorem is likely to remain the most interesting result in the countable case.

§4. Initial segments of $\boldsymbol{D}$ $(\leqq \mathbf{0}^{(1)})$

This has been the writer's primary interest in the theory of initial segments. As we have mentioned, the first result in this area was Sacks' theorem that there is a minimal degree $\leqq \mathbf{0}^{(1)}$; since no r.e. degree is minimal, it follows that in fact there is a minimal degree $<\mathbf{0}^{(1)}$. (This was not the first example of a non-r.e. degree $<\mathbf{0}^{(1)}$, since such a degree had been found earlier by Shoenfield [12] who proved the existence of a degree $< \mathbf{0}^{(1)}$ with no r.e. predecessor other than $\mathbf{0}$. In [19] we extended Shoenfield's method to show that in fact there is a degree $< \mathbf{0}^{(1)}$ which is incomparable with every r.e. degree other than $\mathbf{0}$ and $\mathbf{0}^{(1)}$.) In view of Sacks' result and bearing in mind the conjecture of his that we discussed in §2, it is natural to propose:

C.1. Every finite lattice is $\overset{*}{\Rightarrow} \boldsymbol{D}(\leqq \mathbf{0}^{(1)})$.

(As above this would immediately imply the same result with $\boldsymbol{D}(\leqq \mathbf{0}^{(1)})$ replaced by $\boldsymbol{D}(< \mathbf{0}^{(1)})$.) This conjecture is not of too much interest at the present time, since an affirmative answer is obviously going to be at least as hard to obtain as an affirmative answer to Sacks' conjecture and it is likely (although it does not immediately follow) that this would also be true of a negative answer.

In the countable case, however, we are forced to rephrase the problem, since only countably many topped p.o. sets are $\overset{*}{\Rightarrow} \boldsymbol{D}(\leqq \mathbf{0}^{(1)})$. (This is in contrast with the situation for $\Rightarrow$, since Sacks [10, §4] has shown that any countable p.o. set is $\Rightarrow \boldsymbol{D}(\leqq \mathbf{0}^{(1)})$. In order to rephrase the appropriate conjectures, we need the following definition.

Definition 4.1. For any degree $\boldsymbol{a}$, we say that a p.o. set $(P, \leqq)$ is of degree $\leqq \boldsymbol{a}$ if it is order-isomorphic to a partial ordering $(N, \leqq_B)$ in which $\leqq_B$ is given by a binary relation B (over the integers) of degree $\leqq \boldsymbol{a}$.

Before continuing our discussion of initial segments, it is worth making a temporary diversion concerning the above definition. It might be asked whether if two partial orderings $\leqq_A$ and $\leqq_B$ of N are isomorphic then A and B have the same degree. This is, however, known to be false if the relations concerned are well-orderings: an old theorem of Spector [14] implies that there are isomorphic well-orderings of very different degree. In view of this is it natural to move to the other extreme and propose that there is never much correlation between isomorphism and degree. This again, however, is not strictly true as can be seen from the following examples.

Consider any distributive lattice which consists of an ω-chain of 'lines' and 'diamonds' such as that partly illustrated in Fig. 5. Such a lattice encodes a set of integers S in an obvious way: $n \in S$ if the interval (a_n, a_{n+1}) is a 'line' and $n \notin S$ if it is a diamond'.† If the lattice is given as a partial ordering $(N, \leqq_A)$ then let $S(A)$ be the corresponding set of integers. It is easy to see that $S(A)$ is recursive in $A^{(1)}$. Using this it follows that, for any degree $\boldsymbol{a}$, there are distributive lattices of degree $\leqq \boldsymbol{a}^{(2)}$ which are

† This was used in the work of Kripke, Martin, Sacks and later Jensen on the consistency of a Δ^1 non-constructible set of integers.

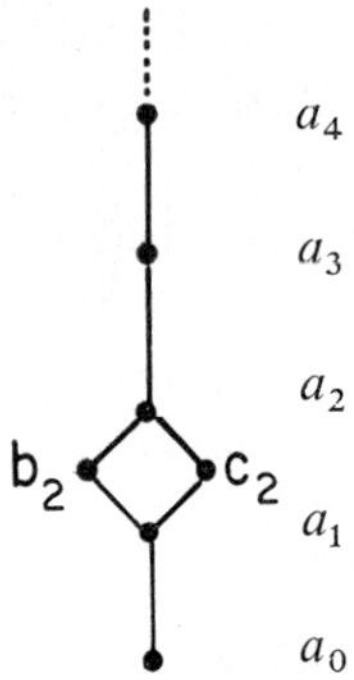

Fig. 5.

not of degree $\leqq \boldsymbol{a}$. This can possibly be strengthened but we are only interested here in showing that there is sometimes a fairly close connection between isomorphism and degree.

To return to initial segments, there are two very natural counterparts to Lachlan's conjecture and the subsidiary conjecture which accompanied it in §3. They are:

C.2. Every bottomed upper semi-lattice of degree $\leqq \mathbf{0}^{(1)}$ is $\overset{*}{\Rightarrow} \boldsymbol{D}(\leqq \mathbf{0}^{(1)})$.

C.3. Every bottomed lattice of degree $\leqq \mathbf{0}^{(1)}$ is $\overset{*}{\Rightarrow} \boldsymbol{D}(\leqq \mathbf{0}^{(1)})$.

There is no evidence that the restriction to p.o. sets of degree $\leqq \mathbf{0}^{(1)}$ is necessary. The only restriction that is immediately apparent is that a p.o. set which is $\overset{*}{\Rightarrow} \boldsymbol{D}(\leqq \mathbf{0}^{(1)})$ has to be of degree $\leqq \mathbf{0}^{(4)}$; this is because relative recursiveness is a 4-quantifier arithmetical relation.† We could therefore extend C.2 and C.3 to two stronger conjectures C.2′ and C.3′ obtained by replacing $\mathbf{0}^{(1)}$ by $\mathbf{0}^{(4)}$ as a restriction on the p.o. sets involved. These conjectures would not trivially reduce to C.2 and C.3, because of the comments that were made after Definition 4.1. In fact, even if C.2 and C.3 are eventually settled affirmatively, we imagine that considerable ingenuity will be required to extend this to C.2′ and C.3′.

It is now time to announce some positive results. In Part IV we shall prove the following constructive counterpart to the Hugill-Lachlan theorem.

THEOREM 4.2. *Every bottomed distributive lattice of degree* $\leqq \mathbf{0}^{(1)}$ *is* $\overset{*}{\Rightarrow} \boldsymbol{D}(\leqq \mathbf{0}^{(1)})$.

The remarks which we made above concerning the replacement of $\mathbf{0}^{(1)}$

---

† L. Feiner has recently proved that there is a linear ordering of degree $\leqq \boldsymbol{Q}^{(5)}$ which is not of degree $\leqq \boldsymbol{Q}^{(4)}$. It easily follows from his result and a trivial generalisation of our Theorem 4.2 that $\boldsymbol{D}(\leqq \boldsymbol{Q}^{(5)})$ is not isomorphic to $\boldsymbol{D}(\leqq \boldsymbol{Q}^{(1)})$. (If these structures are extended to take into account the jump operation then the corresponding weaker negative result is of course trivial.)

by $\mathbf{0}^{(4)}$ are equally relevant here. Since the $\mathbf{0}^{(1)}$ case has been settled, it is worth making the following conjecture:

C.4. Every bottomed distributive lattice of degree $\leqq \mathbf{0}^{(4)}$ is $\stackrel{*}{\Rightarrow} \boldsymbol{D}(\leqq \mathbf{0}^{(1)})$.

All the corollaries of the Hugill-Lachlan theorem have their constructive analogues. It is not worth repeating the full list here, but two of them are of special interest. As we have already mentioned, the first of these was obtained some time ago by the writer using a different method.

COROLLARY 4.3. *There is a nonzero degree* $<\mathbf{0}^{(1)}$ *which has no minimal predecessor.*

The proof of this corollary is exactly the same as that of Corollary 3.5, except that we use Theorem 4.1 instead of Theorem 3.1 and observe that the chain of type $(1, -\omega)$ involved in Corollary 3.5 is recursive.

The other corollary is the most interesting one. Let **EDU** $(\leqq \mathbf{0}^{(1)})$ be the (semantically defined) elementary theory of the degrees $\leqq \mathbf{0}^{(1)}$, formulated exactly as **EDU** was formulated in §3.

COROLLARY 4.4. ***EDL*** *is recursive in* ***EDU*** $(\leqq \mathbf{0}^{(1)})$, *and hence* **EDU** $(\leqq \mathbf{0}^{(1)})$ *is not axiomatisable.*

The proof of this corollary is exactly the same as that of Corollary 3.7, except that instead of using the completeness theorem to assert the existence of a countable model we use the well-known stronger observation that this model can be chosen to be of degree $\leqq \mathbf{0}^{(1)}$ in the obvious way.

Finally, we do not at present know whether Lerman's recent results on non-distributive lattices extend to $\boldsymbol{D}$ $(\leqq \mathbf{0}^{(1)})$ but this seems very probable.

§5. Initial segments of $D(\leqq a)$ for arbitrary recursively enumerable a

This brief section is devoted to part of the theory that will only be properly initiated in the second part of this paper [20]. The first natural question to ask is whether there is a minimal degree below any nonzero r.e. degree. This was proposed by Sacks [10, §12, Q. 1.] and will be answered affirmatively in [20]. The proof is very complicated and it is certain that any stronger theorems will be rather difficult and tedious. We content ourselves here with a conjecture that we do take seriously!

CONJECTURE. *Any bottomed distributive lattice of degree* $\leqq \boldsymbol{a}$ *is* $\stackrel{*}{\Rightarrow} \boldsymbol{D}(\leqq \boldsymbol{a})$ *for every nonzero r.e. degree* $\boldsymbol{a}$.

Needless to say, many other counterparts to conjectures in the previous two sections can be formulated for $\boldsymbol{D}$ $(\leqq \boldsymbol{a})$ with $\boldsymbol{a}$ a nonzero r.e. degree. They seem rather irrelevant until settled in the much less difficult case where $\boldsymbol{a} = \mathbf{0}^{(1)}$.

§6. Uncountable initial segments of $\boldsymbol{D}$

We now swing back to the other extreme. Uncountable initial segments of $\boldsymbol{D}$ certainly exist: $\boldsymbol{D}$ itself is one! A less trivial example is the collection M of all minimal degrees together with $\mathbf{0}$; this is well-known to have cardinality $2^{\aleph_0}$. Also, it has been recently proved by Thomason [16] that if X is a set of cardinality $\leqq 2^{\aleph_0}$ then the distributive lattice of all finite subsets of X is $\overset{*}{\Rightarrow} \boldsymbol{D}$; it had been previously proved by Sacks [10, §3] that this is true if $\overset{*}{\Rightarrow}$ is replaced by $\Rightarrow$. Anyway these examples show that there are a number of initial segments of cardinality $2^{\aleph_0}$. It is clear, however, that any uncountable *linearly*-ordered initial segment of $\boldsymbol{D}$ must have cardinality $\aleph_1$ because $\boldsymbol{D}$ has the c.p.p. Moreover, any uncountable well-ordered initial segment of $\boldsymbol{D}$ must actually be isomorphic to $\aleph_1$. This leads to the first natural question:

U.1. Is $\aleph_1 \overset{*}{\Rightarrow} \boldsymbol{D}$?

It is easy to show using the Axiom of Choice that $\aleph_1 \Rightarrow \boldsymbol{D}$; this application of the Axiom of Choice can be shown to be necessary, but since the structure of $\boldsymbol{D}$ can be so different when the Axiom of Choice is not assumed, it does not seem worth digressing on this topic here. In relation to U.1, it can be proved that:

THEOREM 6.1. *If Set Theory is consistent then it remains consistent on the adjunction of the Axiom of Choice and the statement "*$\aleph_1 \overset{*}{\Rightarrow} \boldsymbol{D}$*".*

The proof uses an extension of Sacks' 'perfect-set forcing' and provides no special problem. It might be suspected that U.1 is completely independent of Set Theory including the Axiom of Choice. At the time of writing this remains an open problem, but we conjecture that this suspicion is correct.† An even more interesting question is whether the Axiom of Constructibility settles U.1 in one way or the other.

It is natural to generalise U.1. Since every initial segment of $\boldsymbol{D}$ is a subsegment of a segment which is an upper semi-lattice, the most general question which this line of enquiry suggests is U.6 below. The other questions are natural steps from U.1 to U.6.

U.2. If $(P, \leqq)$ is a bottomed linearly-ordered set of cardinality $\aleph_1$ having the c.p.p. then $(P, \leqq)$ is $\overset{*}{\Rightarrow} \boldsymbol{D}$.

U.3. If $(P, \leqq)$ is a Boolean algebra of cardinality $\aleph_1$ having the c.p.p. then $(P, \leqq) \overset{*}{\Rightarrow} \boldsymbol{D}$.

† The possibility that U.1 might be completely independent of Set Theory was first voiced convincingly by Professor Jensen at this Colloquium.

U.4. If $(P, \leqq)$ is a bottomed distributive lattice of cardinality $\aleph_1$ having the c.p.p. then $(P, \leqq) \overset{*}{\Rightarrow} \boldsymbol{D}$.

U.5. If $(P, \leqq)$ is a bottomed lattice of cardinality $\aleph_1$ having the c.p.p then $(P, \leqq) \overset{*}{\Rightarrow} \boldsymbol{D}$.

U.6. If $(P, \leqq)$ is a bottomed upper semi-lattice of cardinality $\aleph_1$ having the c.p.p. then $(P, \leqq)$ is $\overset{*}{\Rightarrow} \boldsymbol{D}$.

If U.1 is independent of Set Theory then so are U.2, U.4, U.5 and U.6, and it is hard to imagine how the situation of U.3 could be any different. On the other hand it is fairly certain that the methods used to prove the consistency of U.1 can be extended to prove the consistency of U.4 and hence U.2 and U.3. The consistency of U.5. and U.6 is intimately connected with the truth or falsity of the corresponding propositions in the countable case and is likely to remain open as long as they do.

Finally, although the questions above can be rephrased with $\aleph_1$ replaced by $2^{\aleph_0}$ (with the exception of U.1 where this is irrelevant), there is little point in emphasising the resulting problems until they have been settled in the simpler case where $\overset{*}{\Rightarrow}$ is replaced by $\Rightarrow$.

§7. Some general problems of isomorphism and indiscernibility

We conclude this survey with some problems that have been discussed for some time, although they have not received much attention in the literature because until recently nothing interesting was known about them. They are not problems about initial segments, although at first sight it would appear that very marked progress on initial segments would lead to their solution. We are mentioning them here just in case this is true, but most of our efforts will be directed towards indicating that it is false! The questions that we have in mind are the following.

Q.1. Are $\boldsymbol{D}(\geqq \boldsymbol{a})$ and $\boldsymbol{D}(\geqq \boldsymbol{b})$ indiscernible†, for any two degrees $\boldsymbol{a}$ and $\boldsymbol{b}$?

Q.2. Are $\boldsymbol{D}(\geqq \boldsymbol{a})$ and $\boldsymbol{D}(\geqq \boldsymbol{b})$ order-isomorphic, for any two degrees $\boldsymbol{a}$ and $\boldsymbol{b}$?

The obvious approach to a positive solution in either case would need some sort of very general theorem (much more general than Lachlan's *conjecture*) about embedding initial segments. For example, it might be conjectured that: if $(I, \leqq)$ and $(I', \leqq')$ are two countable bottomed upper semi-lattices such that $(I, \leqq) \overset{*}{\Rightarrow} (I', \leqq')$ and f is an order-isomorphism of $(I, \leqq)$ onto an initial segment of $\boldsymbol{D}$ then f can be extended to an order-

† We prefer this term, which is used in at least one widely-used text-book, to the term 'elementary equivalence' which is prevalent in the literature.

isomorphism f' of $(I', \leqq')$ onto an initial segment of $\boldsymbol{D}$. *Unfortunately, this conjecture is false.* For, if $(I, \leqq)$ is $\boldsymbol{D}(\leqq \boldsymbol{0}^{(1)})$ with the usual ordering and $(I', \leqq')$ is the extended structure formed by adding a single element above $\boldsymbol{0}^{(1)}$ then this conjecture would imply that there is an order-isomorphism f' which maps $\boldsymbol{D}(\leqq \boldsymbol{0}^{(1)})$ onto itself and the new element a of I' onto a degree $\boldsymbol{a} > \boldsymbol{0}^{(1)}$. But then $\boldsymbol{a}$ would be the jump of a lower degree (by a theorem of Friedberg which can be found in [10, §6]). This immediately implies that $\boldsymbol{a}$ is the l.u.b. of two lower degrees $\boldsymbol{a}_1$ and $\boldsymbol{a}_2$ (by a theorem of Sacks [10, §5]). These degrees cannot be both $\leqq \boldsymbol{0}^{(1)}$ and so $(I', \leqq')$ is *not* mapped by f' onto an initial segment of $\boldsymbol{D}$. This shows the falsity of the conjecture above, and no replacement is foreseeable that might help with Q.1. and Q.2.

Further evidence that Q.1 and Q.2 are independent of the usual axioms for Set Theory is that the only positive result known (due to Martin) involves a new Set-Theoretic axiom. Since it is a rather curious and pleasing result, we shall sketch a proof of it here. The new axiom involved is the Axiom of Projective Determinateness. This axiom has recently been used by Addison-Moschovakis [1] and Martin [7] to settle some old problems about the projective hierarchy in descriptive set theory.† We do not need to discuss the axiom itself here but only the following consequence of it. Martin's method (used in [7] as well as for the present theorem) involves first showing that the filter

$$M = \{X \subseteq \boldsymbol{D}\colon (\exists \boldsymbol{a})(\forall \boldsymbol{b})(\boldsymbol{b} \geqq \boldsymbol{a} \rightarrow \boldsymbol{b} \in X)\}$$

restricted to projective sets of degrees (obtained from projective sets of reals in an obvious way) is in fact an ultrafilter, i.e. if X is projective then either X or $\boldsymbol{D} - X$ belongs to M. Also, it is easy to see that M is closed under countable intersections; the resulting intersection may not be projective even though the components are projective, but that is irrelevant for the present application.

THEOREM 7.1 (Martin). *The Axiom of Projective Determinateness implies that there is a degree* $\boldsymbol{a}$ *such that* $\boldsymbol{D}(\geqq \boldsymbol{a})$ *and* $\boldsymbol{D}(\geqq \boldsymbol{b})$ *are indiscernible for all* $\boldsymbol{b} \geqq \boldsymbol{a}$.

Proof. Let $\boldsymbol{T}$ be the subset of the sentences of $\mathscr{L}_{\leqq}$ (used in §3) which is defined by

$$\Phi \in \boldsymbol{T} \;\leftrightarrow\; (\exists \boldsymbol{a})(\forall \boldsymbol{b})(\boldsymbol{b} \geqq \boldsymbol{a} \rightarrow \boldsymbol{D}(\geqq \boldsymbol{b}) \models \Phi).$$

Then the Axiom of Projective Determinateness implies that $\boldsymbol{T}$ is a complete consistent theory, since for each Φ of $\mathscr{L}_{\leqq}$ either $\{\boldsymbol{b}\colon \boldsymbol{D}(\geqq \boldsymbol{b}) \models \Phi\}$ or

† Addison-Moschovakis [1] actually formulates a more general axiom but it is the the axiom above which plays the crucial role.

$\{b : D(\geqq b) \models \neg\Phi\}$ belongs to M (but not both). Since M is closed under countable intersections, it follows that

$$\{b : (\forall\Phi)(\Phi \in T \Rightarrow D(\geqq b) \models \Phi)\}$$

belongs to M. Hence there is a degree a such that, for every degree $b \geqq a$, the set of sentences of $\mathscr{L}_{\leqq}$ valid in $D(\geqq b)$ is T. ∎

We believe that Q.1 and Q.2 are independent of Set Theory and that they may be settled by some new axioms. But at present there is so little evidence that it is difficult to feel very strongly about the problem. It seems, however, an appropriate place to conclude a survey that has been directed towards exhibiting some of the connections between Set Theory and the degrees of unsolvability.

REFERENCES

[1] J. W. ADDISON *and* Y. N. MOSCHOVAKIS, *Some consequences of the axiom of definable determinateness*, Proc. Nat. Acad. Sci. (U.S.A.) **59** (1968), 708–712.

[2] A. GRZEGORCZYK, *Undecidability of some topological theories*, Fund. Math. **38** (1951), 137–152.

[3] D. F. HUGILL, *Initial segments of Turing degrees*, Proc. Lond. Math. Soc. **19** (1969), 1–16.

[4] D. F. HUGILL and A. H. LACHLAN, *Distributive initial segments of the degrees of unsolvability*, Zeits. für Math. Logik und Grundl. der Math. **14** (1968), 457–472.

[5] M. LERMAN, *Some non-distributive lattices as initial segments of the degrees of unsolvability*, Jour. Symb. Logic **34** (1969), 85–98.

[6] D. A. MARTIN, *Category, measure and the degrees of unsolvability* (to appear).

[7] D. A. MARTIN, *The axiom of determinateness and reduction principles in the analytical hierarchy*, Bull. Amer. Math. Soc. **74** (1968), 687–689.

[8] H. ROGERS, *Theory of recursive functions and effective computability* (McGraw–Hill 1967).

[9] G. E. SACKS, *A minimal degree less than* $\mathbf{0}^{(1)}$, Bull. Amer. Math. Soc. **67** (1961), 416–419.

[10] G. E. SACKS, *Degrees of unsolvability*, Ann. of Math. Study no. 55 ((Princeton 1963).

[11] G. E. SACKS, *Measure-theoretic uniformity in recursion theory and set theory*, *Foundations of Mathematics* (Gödel's 60th Birthday Volume), (Springer, 1968.)

[12] J. R. SHOENFIELD, *On the degrees of unsolvability*, Ann. of Math. **69** (1959), 644–653.

[13] J. F. SHOENFIELD, *A theorem on minimal degrees*, Jour. Symb. Logic **31** (1966), 539–544.

[14] C. SPECTOR, *On recursive well-orderings*, Jour. Symb. Logic **20** (1955), 151–163.

[15] C. SPECTOR, *On degrees of recursive unsolvability*, Ann. of Math. **64** (1956), 581–592.

[16] S. K. THOMASON, *A theorem on initial segments of degrees*, Jour. Symb. Logic (to appear).

[17] D. TITGEMEYER, *Untersuchungen über die Struktur des Kleene-Postschen Halbverbandes der Grade der rekursiven Unlösbarkeit*, Doctoral Thesis, Westfälische Wilhelms-universität zu Münster, 1962.

[18] C. E. M. YATES, *Density and incomparability in the degrees less than* $\mathbf{0}^{(1)}$, Jour. Symb. Logic **31** (1966), 301–302.

[19] C. E. M. YATES, *Recursively enumerable degrees and the degrees less than* $\mathbf{0}^{(1)}$, *Set, Models and Recursion Theory* (North-Holland, 1967), 264–271.

[20] C. E. M. YATES, *Initial segments of the degrees of unsolvability*, Part II: *Minimal degrees*, Jour. Symb. Logic (to appear).

SOME APPLICATIONS OF ALMOST DISJOINT SETS

R. B. JENSEN AND R. M. SOLOVAY*

Seminar für Logik, Bonn *University of California, Berkeley.*

1. Introduction. Shoenfield has shown, in [7], that every Σ_2^1 set of integers is constructible ([1]). Shoenfield's proof of this suggests strongly that Σ_2^1 is best possible.

If one assumes the existence of a measurable cardinal, one can completely describe the situation (cf. [9]). It turns out that every constructible set of integers is Δ_3^1. Moreover, there is a certain Δ_3^1 set of integers, $O^{\#}$, in which every constructible real is recursive. ($O^{\#}$ actually satisfies a stronger property than being Δ_3^1. It is the unique solution of a certain Π_2^1 predicate.)

Thus, for the believer in measurable cardinals, it is clear that Shoenfield's theorem is best possible. In the present paper, we show, assuming only the consistency of ZFC ([2]), that there are models of ZFC in which the situation described above obtains. That is, there is a non-constructible set of integers, a, which is a Π_2^1 singleton, and in which every constructible set of integers is recursive.

The first consistency result of this type is due to Kripke and Martin. Using the minimal degree construction of Sacks, they constructed a model of ZF, in which there was a non-constructible Δ_4^1 real; the axiom of choice did not hold in their model. More recently, Jensen [3] has obtained a model in which there is a non-constructible Π_2^1 singleton of minimal degree of non-constructibility.

Our proof uses a certain technique for constructing, in certain Cohen extensions, subsets of ω which encode a great deal of information. This method will be exposed in §2. In §3, we will use this technique to prove the following result:

Let M be a countable transitive model of $\mathrm{ZFC} + V = L$. Let ϑ be an inaccessible cardinal in M. Then there is a set of integers, x, such that $M[x]$ is a model of ZFC. Moreover,

$$\aleph_1^{M[x]} = \vartheta.$$

---

* The second author is a Sloan Foundation fellow. His portion of this work was partially supported by NSF Contract GP–8746.

([1]) For the theory of constructible sets, see [2].

([2]) ZFC (resp. ZF) is Zermelo-Fraenkel set theory with (resp. without) the axiom of choice.

Finally, the cardinals of M which are greater than ϑ remain cardinals in $M[x]$.

In §4, we use the method of almost disjoint sets to construct models in which there are non-constructible Δ_3^1 reals. Finally, in §5, we prove the result stated above by modifying the proof given in §4. We close the paper with a brief discussion of open problems.

2. Almost disjoint sets

2.1. We are going to develop in this section a method, due to Solovay, of producing subsets of ω in Cohen extensions which encode a great deal of information. The method was originally devised to prove the following theorem:

THEOREM. *Let M be a countable transitive model of* ZFC. *We suppose that $\aleph_1^L = \aleph_1$, in M. Let $A \subseteq \aleph_1$, $A \in M$*(3)*. Then there is a Cohen extension of M, $M[x]$, obtained by adjoining the set of integers x such that A is constructible from x and M and $M[x]$ have the same cardinals.*

This theorem will be applied in [4]. For the moment, we use it as a means of motivating the technique of almost disjoint sets.

2.2. We wish to construct a real x that encodes the set A. Our plan is to first construct a sequence, $\{Q_\alpha(\ ), \alpha < \aleph_1\}$ of properties that are suitably independent of each other, and then to choose x so that

$$A = \{\alpha \mid Q_\alpha(x)\}.$$

How do we manufacture this family of properties? We start with the relation $R(x, y)$: $x \cap y$ have infinite intersection. As y ranges over the constructible subsets of ω, $R(\ , y)$ produces $\aleph_1$ distinct properties.

2.3. Speaking heuristically, when should the properties $R(\ , y)$ and $R(\ , z)$ be independent? Clearly, we should require that y and z are infinite. A moment's reflection will make it plausible that $R(\ , y)$ and $R(\ , z)$ are independent if y and z are infinite but $y \cap z$ is finite. If this happens, we say that y and z are *almost disjoint*.

Both the following lemma and its proof are important for us. (The proof shows that almost disjoint families can be constructed in an explicit natural way.) The lemma is well-known.

LEMMA. *There is a pairwise almost disjoint family, $\mathscr{F}$, of subsets of ω of power $2^{\aleph_0}$.*

---

(3) We frequently use set-theoretical concepts relative to a model without explicit mention. Here for example, A is, in M, a set of countable ordinals.

Proof. Let $\{s_i\}$ be some fixed recursive enumeration of the finite sequences of integers. If f is a function mapping ω into ω, let $S(f)$ be $\{j : s_j$ is an initial segment of $f\}$. Then $S(f)$ is infinite.

Now let f, g map ω into ω. Suppose that $f(n) \neq g(n)$. Then if $j \in S(f) \cap S(g)$, s_j has length $\leqq n$. Hence $S(f) \cap S(g)$ is finite, and $S(f)$ and $S(g)$ are almost disjoint. Thus it suffices to take

$$\mathscr{F} = \{S(f) : f \in \omega^{\omega}\}.$$

2.4. The following lemma will allow us to produce the subsets of ω needed to prove Theorem 2.1.

LEMMA. *Let M be a countable transitive model of* ZFC. *Let $Y \in M$ be a subset of $P(\omega)$* (4). *Then there is a subset, x, of ω such that:*

1. *$M[x]$ is a model of* ZFC *with the same cardinals as M.*
2. *If $y \in Y$, $y \cap x$ is finite.*
3. *Suppose $y \in P(\omega)_M$, and $y \cap x$ is finite. Then $y \subseteq y_1 \cup \cdots \cup y_n \cup F$ where $y_i \in Y$ for $i = 1, \ldots, n$, and F is finite.*

(The net effect of clause 3 is to ensure that for $y \in M$, $y \cap x$ is finite only if this is required by clause 2.)

2.5. Our proof of Lemma 2.4 will use "forcing". Our basic reference on forcing will be Shoenfield [6]. However, our terminology differs from Shoenfield's in the following minor respects:

1. If P is the set of conditions, we will choose our ordering on P so that $p \geqq p'$ if p gives more information than p'. (Shoenfield uses the opposite convention.)
2. We say that G is a generic filter on P if it is a generic subset of P in the sense of [6].

We define a partially ordered set, P_Y, as follows. An element p of P_Y is an ordered pair $\langle s, t\rangle$ where s is a finite subset of ω, and t is a finite subset of Y. We say that $\langle s, t\rangle \leqq \langle s', t'\rangle$ if

1) $s \subseteq s'$; 2) $t \subseteq t'$; 3) $s \cap A = s' \cap A$, for any $A \in t$.

The motivation for this definition is as follows: We think of the elements of P_Y as conditions on a "new" set of integers x. If $p = \langle s, t\rangle$, then p tells us that 1) $s \subseteq x$ and 2) $x \cap A = s \cap A$, for each $A \in t$.

Now let G be an M-generic filter on P_Y. Let

$$x = \{n \mid \exists \langle s, t\rangle \in G \text{ with } n \in s\}.$$

We have, clearly, $M \subseteq M[x] \subseteq M[G]$. Thus our assertion that M and $M[x]$ have the same cardinals follows from the Corollary to Lemma 10.2 of [6] and the following lemma. → So $M[x]$ is not a generic extension by the poset P_Y.

(4) $P(\omega)$ is the power set of ω, and ω is the set of non-negative integers.

LEMMA 1. *P_Y satisfies the $\aleph_1$-chain condition.*

Proof. Let $p_1 = \langle s, t_1 \rangle$ and $p_2 = \langle s, t_2 \rangle$ be elements of P_Y with the same first component s. Then p_1 and p_2 are compatible since they have the common extension $\langle p, t_1 \cup t_2 \rangle$.

Thus if $X \in M$ is a pairwise incompatible subset of P_Y, then distinct elements of X must have distinct first components. Since there are only countably many possibilities for the first component, X must be at most countable.

LEMMA 2. *Let $A \in Y$. Then $x \cap A$ is finite.*

Proof. Since G is generic, there is an $\langle s, t \rangle \in G$ with $A \in t$. But then $x \cap A = s \cap A$.

LEMMA 3. *Let $A \in M$. If $x \cap A$ is finite, then there is a finite set s and a finite subset t of Y with $A \subseteq s \cup (\bigcup t)$.*

Proof. Say $A \cap x = F$. Let $\langle s, t \rangle \in G$ force "$A \cap x = F$". Then

$$A \subseteq (F \cup s) \cup (\bigcup t).$$

(Otherwise, pick $n \in A - (F \cup s \cup \bigcup t)$. Let $s' = \{n\} \cup s$. Then $\langle s, t \rangle \leqq \langle s', t \rangle$, so $\langle s', t \rangle$ forces "$n \in x \cap A$ and $x \cap A \subseteq F$", which is absurd.)

Clearly Lemmas 1 through 3 establish Lemma 2.4.

2.6. It is now easy to complete the proof of Theorem 2.1. Let $\{f_\alpha, \alpha < \aleph_1\}$ be an enumeration of the constructible maps of ω into ω, in order of construction. Let $y_\alpha = S(f_\alpha)$ (cf. §2.3). Let $Y = \{y_\alpha : \alpha \in A\}$. Let x be a subset of ω provided by Lemma 2.4. Using clauses 2) and 3) of Lemma 2.4 *and the fact that* $\{y_\alpha, \alpha < \aleph_1\}$ is pairwise almost disjoint, we see that $A = \{\alpha \mid y_\alpha \cap x \text{ is finite}\}$. Theorem 2.1 is proved.

3. Cardinal collapsing with reals

3.1. This section and sections 4 and 5 are independent of one another. Before stating precisely the main result of this section, we review some terminology.

An infinite cardinal, ϑ, is *regular* if for every ordinal $\alpha < \vartheta$, and every map $f: \alpha \to \vartheta$, there is a $\beta < \vartheta$, with range $(f) \subseteq \beta$. If γ is an infinite cardinal, γ^+ is the least cardinal greater than γ. Cardinals of the form γ^+ are called successor cardinals. They are regular. Uncountable regular cardinals which are not successors are *weakly inaccessible*. An infinite cardinal, ϑ, is *strongly inaccessible*, if it is regular, uncountable, and for any cardinal $\lambda < \vartheta$, we have $2^\lambda < \vartheta$.

Let α be an ordinal less than or equal to ϑ. A strictly increasing map, $f:\alpha \to \vartheta$, is continuous if $f(\lambda) = \sup\{f(\xi): \xi < \lambda\}$ whenever λ is a limit ordinal less than α. A map $f: \vartheta \to \vartheta$ is *normal* if it is strictly increasing and continuous. A regular cardinal, ϑ, is *weakly Mahlo* if every normal function $f: \vartheta \to \vartheta$ contains a regular cardinal, λ, in its range (i.e., $\lambda = f(\xi)$, for some $\xi < \vartheta$).

3.2. We now state our precise result. It is more general than the result stated in §1, but is also somewhat more technical.

THEOREM. *Let M be a countable transitive model of* ZFC. *Let ϑ be an uncountable regular cardinal in M. Then there is a real x such that:* 1) *$M[x]$ is a model of* ZFC; 2) $\aleph_1^{M[x]} = \vartheta$; 3) *if ϑ is either not weakly Mahlo or is strongly inaccessible, then every cardinal λ of M, greater than ϑ, remains a cardinal in $M[x]$.*

REMARKS 1) If GCH holds in M, then every uncountable regular cardinal is either a successor or strongly inaccessible, so 3) always applies. Thus our result includes the result stated in §1.

2) If ϑ is a successor cardinal then our claim is trivial. Indeed, let $\vartheta = \gamma^+$, and let $f:\omega \to \gamma$ be a generic collapsing map. (Cf. for example, [8, I. 1.12]. This notion is due, originally, to Levy.) Let x be a real in $M[f]$ which encodes a well-ordering of ω of type γ. The conditions used to adjoin f satisfy the γ^+-chain condition. Hence, the cardinals of M greater or equal to ϑ remain cardinals in $M[x]$. Since γ is countable in $M[x]$, $\vartheta = \aleph_1^{M[x]}$, and we are done with this case.

3.3. We may now assume that ϑ is weakly inaccessible in M, since by the remarks above, the case when ϑ is a successor cardinal is trivial. We now make the additional assumption that ϑ is not weakly Mahlo. (We shall later free ourselves of this assumption.)

The following lemma is due to A. Levy. For a proof we refer the reader to [8, I.3].

LEMMA. *There is a set $A \subseteq \vartheta$ with the following properties*: 1) *$M[A]$ is a model of* ZFC; 2) $\aleph_1^{M[A]} = \vartheta$; 3) *for $\alpha > \vartheta$, α is a cardinal of M if and only if α is a cardinal of $M[A]$.*

(Roughly speaking, A is obtained by adjoining, for each non-zero $\alpha < \vartheta$, a generic collapsing map, $f_\alpha: \omega \to \alpha$. Parts 2) and 3) are verified by showing that the relevant set of conditions satisfies the ϑ-chain condition.)

3.4. We work within M. Our present assumption is that ϑ is not weakly Mahlo. Hence there is a normal map $f: \vartheta \to \vartheta$ such that $f(\alpha)$ is singular for any $\alpha < \vartheta$. Let E be a set of ordinals in M, from which every subset of ϑ, lying in M, is constructible. Thus $L[E] \subseteq M$, and if α is an ordinal less than or equal to ϑ, α is regular in M iff α is regular in $L[E]$.

We say that a set of integers, a, codes an ordinal ξ if there is a well-ordering R, of ω of order type ξ such that

$$a = \{2^i 3^j : \langle i,j\rangle \in R\}.$$

We now place ourselves in $M[A]$. We are going to define, by transfinite induction on ξ, a code, a_ξ, for $f(\xi)$. For ξ not a limit ordinal, we simply take a_ξ to be the least code for $f(\xi)$ in some fixed well-ordering of the reals of $M[A]$.

Now let λ be a limit ordinal less than ϑ. Suppose that $\langle a_\alpha, \alpha < \lambda\rangle$ is the λ-sequence of previously selected codes. Let

$$N = L[E, \langle a_\alpha, \alpha < \lambda\rangle].$$

Note that N has a canonical well-ordering.

We show that $f(\lambda)$ is countable in N. First, if $\alpha < \lambda$, a_α lies in N, so $f(\alpha)$ is countable in N. Thus $f(\lambda) = \sup\{f(\alpha) \mid \alpha < \lambda\} \leqq \aleph_1^N$. If $f(\lambda) = \aleph_1^N$, then $f(\lambda)$ is regular in N, hence regular in $L[E]$. But $f(\lambda)$ is singular in $L[E]$ since it is singular in M. Thus $f(\lambda) < \aleph_1^N$.

We now take a_λ to be the least code for $f(\lambda)$ in the canonical well-ordering of N.

3.5. Our plan now is as follows. Using the techniques of §2, we construct a real x which encodes in a very constructive manner a function $f: R \to R$ defined as follows: 1) $f(\phi) = a_0$; 2) $f(a_\alpha) = a_{\alpha+1}$; 3) at other places, the precise definition of f is unimportant. Hence, using x we can reconstruct the sequence $\langle a_\alpha \mid \alpha < \vartheta\rangle$.

We start with some definitions. Let $y \subseteq \omega$, $n \in \omega$. Define $f(n,y)$ as follows: 1) $f(n,y): \omega \to \omega$; 2) $f(n,y)(0) = n$; 3) $f(n,y)(k+1) = 1$ if $k \in y$, and equals 0 otherwise. Let $R(x,y) = \{n \mid S(f(n,y)) \cap x \text{ is infinite}\}$. Using the results of §2, the following lemma is clear.

LEMMA. *There is a real* x *such that* 1) $M[A,x]$ *is a model of* ZFC *with the same cardinals as* $M[A]$; 2) $R(x,\phi) = a_0$; 3) *If* $\xi < \vartheta$, $R(x, a_\xi) = a_{\xi+1}$.

3.6. *Lemma. The sequence* $\langle a_\xi \mid \xi < \vartheta\rangle$ *is constructible from the pair* $\langle E, x\rangle$.

Proof. We work in $L[E,x]$.

We define a_ξ by transfinite induction on ξ as follows: 1) $a_0 = R(x,0)$; 2) $a_{\xi+1} = R(x,a_\xi)$. 3) If λ is a limit ordinal, and $\langle a_\alpha \mid \alpha < \lambda\rangle$ has already been defined, let γ be the sup of ordinals coded by elements of $\{a_\alpha \mid \alpha < \lambda\}$. (Thus $\gamma = f(\lambda)$.) Let a_λ be the least code for γ in order of construction from $\langle E, \langle a_\alpha, \alpha < \lambda\rangle\rangle$. The lemma is now clear.

We know that ϑ is a cardinal in $M[A, x]$ and hence in $M[x]$. On the other hand, if $\alpha < \vartheta$, then $a_\alpha \in M[x]$, and hence α is countable in $M[x]$. Hence $\vartheta = \aleph_1^{M[x]}$. We also know that if α is a cardinal of M greater than ϑ, then α is a cardinal in $M[A,x]$ and a fortiori, in $M[x]$. Thus the special case when ϑ is not weakly Mahlo is now disposed of.

3.7. The general case will be reduced to the special case by means of the following lemma.

LEMMA. *Let M be a countable transitive model of* ZFC, *ϑ a weakly Mahlo cardinal. Then there is a map $f: \vartheta \to \vartheta$ such that*

1. *$M[f]$ is a model of* ZFC.
2. *ϑ is weakly inaccessible in $M[f]$.*
3. *If ϑ is strongly inaccessible in M, M and $M[f]$ have the same cardinals.*
4. *The map f is normal.*
5. *If $\alpha < \vartheta$, then $f(\alpha)$ is singular in M.*

Let us first see that the lemma allows us to prove Theorem 3.2. We have only the case when δ is weakly Mahlo to handle. Let f be given by the present lemma. Then since ϑ is weakly inaccessible in $M[f]$, we can find a subset $A \subseteq \vartheta$ by Lemma 3.3 such that $\vartheta = \aleph_1$ in $M[f, A]$. We can now construct x as above since $f(\alpha)$ is singular in M for all $\alpha < \vartheta$.

3.8. We turn to the proof of Lemma 3.7. The function f will be constructed via a "forcing" argument. We first describe the set P of conditions. An element $h \in P$ will be a function mapping some ordinal $\alpha < \vartheta$ into ϑ. We require that h satisfy the following conditions:

1) h is strictly increasing and continuous.

2) If $\beta < \alpha$, $h(\beta)$ is singular.

3) If α is a limit ordinal, then

$$\sup \{h(\beta): \beta < \alpha\}$$

is singular.

(Note that if clause 3) fails, h has no extension to a function $h': \alpha + 1 \to \vartheta$ which satisfies 1) and 2).)

We order P by inclusion.

LEMMA. *Let $h \in P$. Let $\xi < \vartheta$. Then h has an extension h' such that $h'(\xi)$ is defined.*

Proof. Let $\alpha = \text{domain}(h)$. If α is a limit ordinal, we extend h by continuity to a function with domain $\alpha + 1$ (i.e., put $h(\alpha) = \sup\{h(\beta): \beta < \alpha\}$).

So we may assume that α is not a limit ordinal, and in fact that $\alpha = \beta + 1$ for some ordinal β. Put

$$\lambda = \max(h(\beta), \xi).$$

Define $h': \xi + 1 \to \vartheta$ by putting

$$h'(\gamma) = h(\gamma) \text{ if } \gamma \leqq \beta,$$

$$h'(\gamma) = \lambda + \gamma \quad \text{if } \beta < \gamma \leqq \xi.$$

Then h' is clearly strictly increasing and continuous. If $\beta < \gamma \leqq \xi$, $h'(\gamma)$ is not even a cardinal, so it is certainly singular. The lemma is proved.

Let G be an M-generic filter on P. Let $f = \bigcup G$. Using the lemma just proved, it is clear that f is a normal function from ϑ into ϑ, and that for $\alpha < \vartheta$, $f(\alpha)$ is singular in M.

Since f is obtained via forcing, $M[f]$ is a model of ZFC.

3.9. The following lemmas will establish that ϑ is weakly inaccessible in $M[f]$.

LEMMA 1. *Let α be a limit ordinal. Suppose that $h_0 \in P$, that $\alpha \in$ domain h_0, and that $\{h_\xi, \xi < \alpha\}$ is a strictly increasing chain of conditions. Then $\bigcup_{\xi<\alpha} h_\xi$ lies in P.*

Proof. Let $h = \bigcup_{\xi<\alpha} h_\xi$. Let $\gamma =$ domain h. Since ϑ is regular, $\gamma < \vartheta$. It is clear that h is strictly increasing and continuous and that $h(\xi)$ is singular for $\xi < \gamma$. Let $\gamma_1 = \sup\{h(\xi): \xi < \gamma\}$. Then

$$\mathrm{cf}(\gamma_1) \leqq \mathrm{cf}(\gamma) = \alpha < \gamma \leqq \gamma_1.$$

Thus γ_1 is singular, and $h \in P$.

LEMMA 2. *Let $\alpha < \vartheta$. Let $g: \alpha \to M$, $g \in M[f]$. Then $g \in M$.*

Proof. Let $\dot{g}$ be a term denoting g. Let $h \in G$ force that $\dot{g}: \alpha \to M$. We may extend h if necessary so that $\alpha \in$ domain h. Using Lemma 1, we see that for every $h' \geqq h$, there is an $h'' \geqq h'$ which decides all the values of $\dot{g}$. Hence there is an $h'' \in G$ which decides all the values of $\dot{g}$, and $g \in M$. (Cf. the proof of Lemma 10.6 of [6].)

It follows readily from Lemma 2 that if $\alpha < \vartheta$, α has the same subsets in M and $M[f]$. Thus ϑ is a limit cardinal in $M[f]$. Also Lemma 2 implies that ϑ is regular in $M[f]$.

We now know, from the remark of the preceding paragraph that all cardinals of M which are $\leqq \vartheta$ remain cardinals in $M[f]$. If ϑ is strongly inaccessible, P has power ϑ, so no cardinals of M above ϑ are destroyed in $M[f]$. (Cf. Lemma 10.2 of [6].) Lemma 3.7 has been proved.

3.10. The proof of Theorem 3.2 in the special case that ϑ is not Mahlo is due to Solovay. Lemma 3.7 and its use for proving Theorem 3.2 in the case when ϑ is Mahlo is due to Jensen.

4. A non-constructible Δ_3^1 real: consistency

4.1. In this section, we prove a theorem, due to Solovay, which is weaker than the result announced in §1. In §5, we indicate the modifications, due to Jensen, needed to prove the full result stated in §1. We shall prove in this section the following theorem:

THEOREM. *Let M be a countable transitive model of* ZFC + $V = L$. *Then there is a transitive model N of* ZFC *such that*:

1) *For a certain Π_2^1 predicate, $S(x)$, which will be described below, we have*

$$N \models (\exists! x \subseteq \omega) S(x).$$

2) *Let $a \in N$ be the unique subset of ω such that $N \models S(a)$. Then $a \notin M$, and $N = M[a]$.*

3) *a is Δ_3^1. (This is clear since a is a Π_2^1 singleton by* 1) *and* 2). Cf. [9, Lemma 2.9].)

4.2. We proceed to outline the construction of N.

Our proof turns on constructing a one-parameter family, $T_i(x)$, of Π_2^1 predicates, and a sequence of reals, a_i, such that

1) $M[\langle a_i, i \in \omega \rangle] \models T_i(a_i)$.

2) Let $j \in \omega$. Suppose that $a_i' = a_i$ for $i \neq j$, but $a_j' = 0$. Then

$$M[\langle a_i', i \in \omega \rangle] \models \neg (\exists x) T_j(x).$$

Thus the statements $(\exists x) T_j(x)$ are independent of one another (5).

Once we have such a family it turns out to be an easy matter to construct the desired model N as an inner model of $M[\langle a_i, i \in \omega \rangle]$. For a suitable choice of a, we will take $N = M[\{\langle i, a_i \rangle : i \in a\}]$. (This idea of getting definability results by passing to inner models is due to K. McAloon [5].)

4.3. We next wish to indicate the ideas behind the construction of the predicates T_i. We are going to construct a family $\{f_{\alpha,i} : \alpha < \aleph_1, i < \omega\}$ of elements of ω^ω in M. The predicate $T_i(x)$ will then express

$$(\forall \alpha)(x \cap S(f_{\alpha.j}) \text{ is finite} \leftrightarrow j = i).$$

In order to ensure that the T_i's have the independence properties listed above, we will arrange that the $f_{\alpha,i}$'s look very much alike. We will do this by making the $f_{\alpha,i}$'s generic (in essentially the original Cohen sense) over larger and larger countable submodels of M.

4.4. We turn to the details. We begin with some preliminary material needed for the definition of the family $\{f_{\alpha.i}\}$.

---

(5) Note that this is a somewhat different notion of independence than that of 2.2–3.

We place ourselves inside M. We wish to remark first that the transitive ground model needed for forcing need not be a model of ZFC. On the other hand, in ZFC, we can prove the existence of transitive models of various weaker theories. For example, let $\boldsymbol{T}$ be the following theory:

1) All the axioms of ZFC except the power set axiom.
2) $V = L$.
3) Every ordinal is countable.

Let λ be an ordinal. Let L_λ be the first λ sets in the canonical well-ordering of L. Then $L_{\aleph_1} \models \boldsymbol{T}$. Hence, if $V = L$, and x is any hereditarily countable set, there is a countable transitive model, M', of $\boldsymbol{T}$ with $x \in M'$. (For example, let t be the transitive closure of $\{x\}$. Then t is countable, so $t \in L_{\aleph_1}$. Let M'' be a countable elementary submodel of $L_{\aleph_1}$ such that $t \subseteq M''$. Finally let M' be the transitive set ε-isomorphic to M''.) We note that the forcing technique applies just as well for models of $\boldsymbol{T}$ as for models of ZFC.

4.5. We order $\aleph_1 \times \aleph_0$ lexicographically; it is then order isomorphic to $\aleph_1$. Let $\langle \alpha_\xi, n_\xi \rangle$ be the ξth member of $\aleph_1 \times \aleph_0$. In M, we define, by induction on the countable ordinal ξ, the following:

1) A countable transitive model, M_ξ, of $\boldsymbol{T}$.
2) A function f_ξ mapping ω into ω.

Suppose then that f_α is defined for all $\alpha < \xi$. We take M_ξ to be the smallest transitive model of $\boldsymbol{T}$ containing the sequence $\{\langle \alpha, f_\alpha \rangle : \alpha < \xi\}$. This makes sense since $V = L$ is an axiom of $\boldsymbol{T}$ so the transitive models of $\boldsymbol{T}$ are well-ordered by inclusion. Note that M_ξ is countable.

We let P be the partially ordered set appropriate to adding a generic function $f: \omega \to \omega$. Thus P is the set of all finite sequences of non-negative integers. We say that $t \leqq s$ iff t is an initial segment of s. Since M_ξ is countable, there is a map $f: \omega \to \omega$, generic over M_ξ, such that s_{n_ξ} is an initial segment of f. Let f_ξ be the least such f. Clearly $f_\xi \neq f_\alpha$, for any $\alpha < \xi$, since $f_\alpha \in M_\xi$.

We now let $f_{\langle \alpha_\xi, n_\xi \rangle}$ be f_ξ, and define $T_i(x)$ as in §4.3.

LEMMA. *The predicate $T_i(x)$ is Π^1_2 (in the two variables i and x).*

Proof. We introduce the following extension $\boldsymbol{T}'$ of $\boldsymbol{T}$. $\boldsymbol{T}'$ is obtained from $\boldsymbol{T}$ by adding an axiom that every countable ordinal lies in some countable transitive model of $\boldsymbol{T}$. Clearly $L_{\aleph_1} \models \boldsymbol{T}'$, and for every countable ordinal, ξ, ξ lies in a countable transitive model of $\boldsymbol{T}'$. The definition of the sequence $\{f_{\alpha,i}\}$ can be given in any model of $\boldsymbol{T}'$. If M' if a countable transitive model of $\boldsymbol{T}'$, and $\alpha \in M'$, then $f_{\alpha,i}$ is the same if computed in M' as it is in M.

We can code binary relations on ω by subsets of ω in an obvious way. Moreover, if a codes the binary relation R_a on ω, then the property "$(\omega; R_a)$ is isomorphic to a transitive model of $\boldsymbol{T}'$" is a Π^1_1 property of a.

Now $T_i(x)$ holds, by the remarks above, just in case for every countable transitive model, M', of $\boldsymbol{T}'$, we have

$$(\forall \alpha \in M')(x \cap S(f_{\alpha \cdot j}^{M'}) \text{ is finite } \leftrightarrow j = i).$$

By the remarks above concerning the coding of countable transitive models of $\boldsymbol{T}'$, this has the form

$$(\forall a)(S_1(a) \rightarrow S_2(a, x, i))$$

where S_1 is Π_1^1 and S_2 is Δ_1^1. Hence the lemma is clear.

4.6. We are going next to construct the sequence $\{\langle a_i, i < \omega \rangle\}$. The sequence will be obtained by forcing, and our first task is to describe the appropriate set of conditions.

Let $i < \omega$. We let Q_i be the partially ordered set of conditions appropriate for adding a set a_i which satisfies $T_i(a_i)$. (Cf. §2.) Thus an element of Q_i is an ordered pair $\langle s, t \rangle$ where s is a finite subset of ω and t is a finite subset of

$$\{S(f_{\alpha . i}) \colon \alpha < \aleph_1\}.$$

The reader may supply the description of the ordering on Q_i by analogy with §2.2.

Let P be the weak direct product of the Q_i's. Thus an element of P is a unction f with domain ω such that $f(i) \in Q_i$ for all i, and $f(i) = \langle \phi, \phi \rangle$ for all but finitely many i. We order P by saying that $f_1 \leqq f_2$ iff for all i, $f_1(i) \leqq f_2(i)$.

Lemma 1. *P satisfies the $\aleph_1$-chain condition.*

Proof. If $f \in P$, define a map $\pi_1(f)$ with domain ω by letting $\pi_1(f)(n)$ be the first component of $f(n)$. It is clear from the definition of P that $\pi_1(f)$ has only countably many possible values. Moreover, if $\pi_1(f) = \pi_1(g)$, then f and g are compatible. (Cf. the proof of Lemma 1 of §2.2.) The lemma follows.

We now select an M-generic filter, G, on P. Let

$$a_i = \{n \in \omega \mid (\exists f \in G)(f(i) = \langle s, t \rangle \wedge n \in s)\}.$$

The following lemma is clear from the methods of §2.

Lemma 2. *Let $N_1 = M[\langle a_i, i \in \omega \rangle]$. Then for all i, $N_1 \models T_i(a_i)$.*

The following lemma is clear from the definition of the predicates T_i.

Lemma 3. *Let N' be a model of ZFC with $M \subseteq N' \subseteq N_1$. Let $x \in N'$. Then $N_1 \models T_i(x)$ iff $N' \models T_i(x)$. (This result is a special case of a general absoluteness result for Π_2^1 statements due to Shoenfield [7].)*

4.7. We briefly recall the product lemma (cf. [6, §8]). Let M be a transitive model of ZFC, or of the theory $\boldsymbol{T}$ considered above. Let P_1, P_2 be partially ordered sets in M. We partially order $P_1 \times P_2$ by putting $\langle p_1, p_2 \rangle \leqq \langle p'_1, p'_2 \rangle$ iff $p_1 \leqq p'_1$ and $p_2 \leqq p'_2$. The product lemma characterizes the M-generic filters G on $P_1 \times P_2$ as follows: they are precisely the sets of the form $G_1 \times G_2$ with G_1 an M-generic filter on P_1 and G_2 an $M[G_1]$-generic filter on P_2.

The product lemma yields, after an easy induction on n, the following characterization of generic filters on an n-fold cartesian product $P_1 \times \cdots \times P_n$. They have the form $G_1 \times \cdots \times G_n$ where, for each i, G_i is an $M[G_1, \ldots, G_{i-1}]$-generic filter on P_i.

Suppose now that P is fixed and that $G_1 \times \cdots \times G_n$ is an M-generic filter on P^n. Let π be a permutation of $\{1, \cdots, n\}$. Then π induces an automorphism of P^n, in M. It follows (cf. [6, Lemma 9.1]) that $G_{\pi(1)} \times \cdots \times G_{\pi(n)}$ is an M-generic filter on P^n. In particular, if we take π such that $\pi(n) = j$, we see that G_j is generic over $M[G_1, \ldots, G_{j-1}, G_{j+1}, \ldots, G_n]$.

4.8. We are going now to study the model obtained by omitting a_j from N_1. To be precise, let $a'_i = a_i$ for $i \neq j$. Let $a'_j = \phi$. We put $N^j = M[\langle a'_i, i < \omega \rangle]$. (The symbol $\langle a'_i, i < \omega \rangle$ denotes the sequence whose ith member is a'_i.)

Let R_j be the set of $f \in P$ such that $f(j) = \langle \phi, \phi \rangle$. We have a direct product decomposition

$$P = Q_j \times R_j$$

and so the product lemma applies. Let $G_j = G \cap R_j$. Then by the product lemma, G_j is an M-generic filter on R_j. Clearly $N^j \subseteq M[G_j]$.

We shall view N_j as obtained via forcing using R_j as the set of conditions.

THEOREM. *Let $x \in N^j$, $x \subseteq \omega$. Then $N^j \models \neg T_j(x)$.*

Proof. We suppose the contrary and proceed to derive a contradiction. Let $x \in N^j$ be such that $N^j \models T_j(x)$. Let τ be a term denoting x. Using Lemma 1 of §4.6, we may assume that $\tau \in L_{\aleph_1}$. Let $p \in G_j$ force $T_j(\tau)$. Clearly $p \in L_{\aleph_1}$.

We fix an elementary submodel M' of $L_{\aleph_1}$ such that p, τ lie in M', and M' is countable in M.

LEMMA 1. *M' is transitive* (6).

Proof. Let ξ be a non-zero ordinal in M'. Then $\xi \in L_{\aleph_1}$, so ξ is countable. Let $f: \omega \to \xi$ be the least map of ω onto ξ. Then f is definable from ξ in

(6) This lemma merely saves us the work of transitively collapsing M'. It is not really necessary.

$L_{\aleph_1}$, so f lies in the elementary submodel M'. Clearly $\omega \subseteq M'$. It follows that $\xi = \{f(n): n \in \omega\}$ is included in M'. Hence the ordinals of M' form an initial segment of the ordinals. Since $V = L$ holds in M', the lemma follows.

We work within M. Let $\mathscr{F}_n$ be a maximal pairwise incompatible subfamily of R_j which decide the statement "$n \in \tau$" (that is, which force "$n \in \tau$" or its negation). For definiteness, we take $\mathscr{F}_n$ to be the least such family in the canonical well-ordering of L. By Lemma 1 of §4.6, $\mathscr{F}_n$ is countable. Hence $\mathscr{F}_n \in L_{\aleph_1}$. Since $\mathscr{F}_n$ is definable in $L_{\aleph_1}$ from p and τ, we have $\mathscr{F}_n \in M'$.

Lemma 2. *Let $G' \subseteq R_j$ be generic over M. Then $G' \cap \mathscr{F}_n$ is non-empty.*

Proof. It is easily seen that the set

$$X = \{q \in R_j: q' \leqq q \text{ for some } q' \in \mathscr{F}_n\}$$

is dense and lies in M. Hence $G' \cap X \neq \phi$. Let then $q \in G'$, $q' \in \mathscr{F}_n$ with $q \geqq q'$. It follows that $q' \in G'$, which proves the lemma.

We let ξ be the least ordinal not in M'. Since $T_j(x)$ holds in N^j, we have

$$x \cap S(f_{\xi,j}) = F$$

where F is finite. Let $p' \in G_j$ force $\tau \cap S(f_{\xi,j}) = F$. We may assume that $p' \geqq p$.

We say that a function $f_{\lambda,k}$ appears in p' if for some n, s, t, we have $p'(n) = \langle s, t\rangle$ and $S(f_{\lambda,k}) \in t$. It then follows than $n = k$. (Cf. the way P was constructed.) Note also that $f_{\xi,j}$ does not appear in p', since $p' \in R_j$.

Lemma 3. *Let $g_1, \ldots, g_m$ be the set of functions which appear in p' but do not lie in M'. Then $f_{\xi,j}$ is generic over $M'[g_1, \ldots, g_m]$ (with respect to the set of conditions, P, of* §4.5).

Proof. Let $\{h_1, \ldots, h_{m+1}\}$ be $\{g_1, \ldots, g_m, f_{\xi,j}\}$ arranged in their order of construction in §4.5. Then it is clear from the construction of §4.5 that h_j is generic over $M'[h_1, \ldots, h_{j-1}]$ for each j between 1 and $m+1$. Hence the lemma follows from the remarks of §4.7.

Let N' be $M'[g_1, \ldots, g_m]$. Let

$$Y = \{n \in \omega: \neg p' \Vdash n \notin \tau\}.$$

Thus Y is the set of possible members of τ given p'.

Lemma 4. $Y \in N'$.

It follows easily from Lemma 2 that

$$Y = \{n: (\exists q' \in \mathscr{F}_n)\ (q' \text{ is compatible with } p' \text{ and } q' \Vdash n \in \tau\}.$$

We know that the sequence $\{\mathscr{F}_n,\ n \in \omega\}$ lies in M'. The relevant forcing notion is definable in $L_{\aleph_1}$ and hence in M'. Clearly $p' \in N'$. Finally, we can clearly determine in N' whether or not p' and q' are compatible. The lemma follows.

LEMMA 5. $Y \cap S(f_{\xi,j}) \subseteq F$.

Proof. Suppose not. Then there is a $p'' \geqq p'$ and a $n \notin F$ such that $n \in S(f_{\xi,j})$ and $p'' \Vdash n \in \tau$. This contradicts the fact that $p'' \geqq p'$ and p' forces $\tau \cap S(f_{\xi,j}) = F$.

We now exploit the fact that $f_{\xi,j}$ is generic over N'. It follows that there is a finite sequence s such that whenever $f: \omega \to \omega$ extends s, and f is generic over N', then $Y \cap S(f) \subseteq F$.

We choose k so that 1) $f_{\xi,k}$ does not appear in p', 2) $k \neq j$, 3) $f_{\xi,k}$ extends s. (Our construction of $\{f_{\lambda,n}\}$ was arranged so that $f_{\lambda,n}$ extends s_n. Hence we can choose k satisfying 3) simply by taking s_k a large extension of s.) It follows from the product lemma that $f_{\xi,k}$ is generic over N'. (Cf. Lemma 3.) Thus by the remark of the preceding paragraph, we have $Y \cap S(f_{\xi,k}) \subseteq F$.

We have however $T_j(x)$. This means that $x \cap S(f_{\xi,k})$ is infinite. Select $n \in X \cap S(f_{\xi,k})$, with $n \notin F$. There is a $p'' \in G_j$ with $p'' \Vdash n \in \tau$. We may as well assume $p'' \geqq p'$, since $p' \in G_j$. Hence $n \in Y$. But this contradicts the fact that $Y \cap S(f_{\xi,k}) \subseteq F$. The theorem is now proved.

4.9. It is now easy to complete the proof of Theorem 4.1. As we mentioned earlier, we are going to form N as an inner model of N_1. We begin by defining a certain set of integers, a, inside N_1. The final model N, will then be $M[a]$.

Let $\{s_j\}$ be a recursive enumeration without repetitions of the set of sequences of non-negative integers. We choose $\{s_j\}$ so that whenever s' is an initial segment of s, s' precedes s in the sequence $\{s_j\}$. Thus s_0 is the void sequence.

Our construction of a will be arranged so that for each $j \in \omega$, exactly one of the two numbers $2j$, $2j+1$ will lie in a. We put $b_j = a_{2j}$ if $2j \in a$, and $b_j = a_{2j+1}$ if $2j+1 \in a$. Thus b_j will be defined as soon as we decide which of $2j$, $2j+1$ lies in a.

We now decide, by induction on j, which of $2j$, $2j+1$ will lie in a. To start things off, we put 0 in a. Now suppose that $j > 0$. Then s_j is a sequence of length $l > 0$. Let s_k be the initial subsequence of s_j of length $l-1$. (Thus $k < j$, and b_k is defined.) Let n_j be the last member of s_j. *We put* $2j$ *in* a

iff $n_j \in b_k$; otherwise we put $2j+1$ in a. This completes our inductive definition of a.

LEMMA 1. *$M[a] \models (\exists x)T_j(x)$ if and only if $j \in a$.*

Proof. Suppose first that $j \in a$. Then a_j will be equal to b_k, where k is the integral part of $j/2$. Now it is clear from our construction that b_k is recursive in a. (To find out if $n \in b_k$, compute the j such that s_j is the concatenated sequence $s_k^\frown n$ and see whether $2j \in a$.) Thus $a_j \in M[a]$. But by Lemma 3 of §4.6,

$$M[a] \models T_j(a_j).$$

Suppose conversely that $j \notin a$. Then a_j is not of the form b_k for any k. Thus the inductive definition of a can be carried out in N^j, and so $M[a] \subseteq N^j$. Theorem 4.8 and Lemma 4.6.3 now show that $M[a] \models \neg(\exists x)T_j(x)$, as desired.

It follows from the proof of this lemma that $a \notin M$. Indeed, a_0 is recursive in a, and it is clear from Theorem 4.8 that $a_0 \notin M$.

The following lemma will complete the proof of Theorem 4.1.

LEMMA 2. *In $M[a]$, a is the unique solution of a Π^1_2 predicate.*

Proof. We first note that the proof of Lemma 1 can be used to construct a primitive recursive function f such that if $j \in a$, then $n \in a_j$ iff $f(j,n) \in a$. It is clear that a has the following two properties:

α) If $j \geqq 0$ then exactly one of the integers $2j$, $2j+1$ lies in a.

β) If $j \in a$, and if $c = \{n \mid f(j,n) \in a\}$, then $T_j(c)$.

Using Lemma 4.5, we see that the conjunction of α) and β) asserts that a satisfies a certain Π^1_2 property, S.

To complete the proof of the lemma, we show that if $b \in M[a]$, and $M[a] \models S(b)$, then $b = a$. Indeed by clause β of S, and Lemma 1 we see that $b \subseteq a$. Since a and b both satisfy clause α of S and $b \subseteq a$, we have $b = a$. The lemma is proved.

5. A Δ^1_3 recursive upper bound for L.

5.1. In this section, we are going to prove the following:

THEOREM. *Suppose that* ZFC *has countable transitive models. Then there is a set of integers, a, and a countable transitive model of* ZFC, *N, such that*

1) $a \in N$.

The following propositions hold in N:

2) *Every set is constructible from a.*

3) *For a certain* Π_2^1 *predicate,* S, *we have* $(\forall x \subseteq \omega)(S(x) \leftrightarrow x = a)$. *(It follows from this that* a *is* Δ_3^1.)

4) *If* $y \subseteq \omega$ *is constructible, then* y *is recursive in* a. *(It follows from this that* a *is not constructible. For example, if* a *is constructible, so is its recursive jump,* a', *and* a' *is known not to be recursive in* a.)

Our proof of this theorem will be closely modeled on the proof of Theorem 4.1. Because of this, we omit many details. The reader should have no trouble filling in these details using the proof of Theorem 4.1 as a model. In order to make the analogy as transparent as possible, we use the same symbols for the Theorem 5.1 analogs of various objects as we did for their Theorem 4.1 prototypes.

5.2. We start with a countable transitive model M of $\mathrm{ZFC} + V = L$. We are going to construct first an outer model N_1. We shall arrange that in N_1, $\aleph_1^L$ is countable. We shall also add a sequence, g_i, of maps of $\aleph_1^L$ into itself. Because $\aleph_1^L$ is countable, subsets of $\aleph_1^L$ generate reals. We shall again construct a family $\{T_i(x)\}$ of "independent" Π_2^1 predicates. Roughly speaking, g_i will insure that $(\exists x)T_i(x)$ is valid. Our final model, N, will be an inner model of N_1. However, because $\aleph_1^L$ is countable in N_1, we will be able to pick N so that clause 4) holds.

5.3. The following theory $\boldsymbol{T}$, is the analog of the theory, $\boldsymbol{T}$, of 4.4:

1) All the axioms of ZFC except the power set axiom.
2) $V = L$.
3) The power set of ω exists. (It follows that $\aleph_1$ exists.)
4) Every ordinal has cardinality $\leqq \aleph_1$.

We now discuss, inside M, the situation as to models of $\boldsymbol{T}$. (Since we are in M, $V = L$.) Clearly $L_{\aleph_2} \models \boldsymbol{T}$. Moreover if x is a set hereditarily of power $\leqq \aleph_1$, then there is an ordinal λ with $\aleph_1 < \lambda < \aleph_2$ such that L_λ is a model of $\boldsymbol{T}$ containing x.

We note that in §4 an important role was played by countable models of $\boldsymbol{T}$. Here the important models of $\boldsymbol{T}$ will have the form L_λ with $\aleph_1 < \lambda < \aleph_2$.

5.4. Instead of generic maps from ω to ω, we will consider generic maps of $\aleph_1^L$ into $\aleph_1^L$. The appropriate set of conditions, P, will be the set of functions f with domain a countable ordinal, and range a subset of $\aleph_1$. We order P by inclusion.

LEMMA. *We assume* $V = L$. *Let* λ *be an ordinal between* $\aleph_1$ *and* $\aleph_2$ *such that* L_λ *is a model of* $\boldsymbol{T}$. *Then* $P \in L_\lambda$. *If* $p \in P$, *there is an* L_λ*-generic filter* G *on* P, *such that* $p \in G$.

Proof. Clearly $P \subseteq L_{\aleph_1} \subset L_\lambda$. Since $P(\omega)$ is a set in L_λ and the replacement axiom is valid in L_λ, we have $P \in L_\lambda$.

We can enumerate the dense subsets of P lying in L_λ in a sequence of length $\aleph_1$. Since P is closed under countable increasing unions, the construction of G is straightforward. (Cf. [6, proof of Lemma 10.5].)

We say, of course, that $f: \aleph_1^L \to \aleph_1^L$ is generic over L_λ if $f = \bigcup G$ for some L_λ-generic filter, G, on P.

5.5. We can now carry over the material of §4.5. First, let $\{p_\xi, \xi < \aleph_1\}$ be the enumeration of P in order of construction. Let $\aleph_2 \times \aleph_1$ be lexicographically ordered.

In imitation of §4.5, we construct a chain, $\{M_\xi, \xi < \aleph_2\}$, of models of $\boldsymbol{T}$ and a family of functions $\{f_{\alpha,\beta}: \langle \alpha, \beta \rangle \in \aleph_2 \times \aleph_1\}$. In particular, we have:

1) M_ξ is a transitive model of $\boldsymbol{T}$ such that $\aleph_1 < OR^{M_\xi} < \aleph_2$.

2) Let $\langle \alpha, \beta \rangle$ be the ξth member of $\aleph_2 \times \aleph_1$. Then $f_{\alpha,\beta}$ is a map of $\aleph_1^L$ into $\aleph_1^L$ generic over M_ξ, which extends p_β.

(Of course, the entire construction takes place inside the model M.)

5.6. We are now going to adapt the "almost disjoint set" formalism to fit our present needs. Let g_1, g_2 be maps of $\aleph_1^L \to \aleph_1^L$. We say that g_1 and g_2 are *almost disjoint* if

$$\operatorname{lub}\{\alpha \mid g_1(\alpha) = g_2(\alpha)\} < \aleph_1^L.$$

We shall need a way of manufacturing almost-disjoint families. Let $h: \aleph_1^L \to \aleph_1^L$ be constructible. We define a function

$$S(h):\ \aleph_1^L \to \aleph_1^L$$

as follows: Let $\alpha < \aleph_1$. Let $f \mid \alpha$ be the restriction of f to α. Say $f \mid \alpha$ is the ξth member of $L_{\aleph_1}$ in the canonical well-ordering of $L_{\aleph_1}$ (by order of construction). Then we put

$$S(f)(\alpha) = \xi.$$

It is clear that if f and g are distinct constructible maps of $\aleph_1^L$ into $\aleph_1^L$. then $S(f)$ and $S(g)$ are almost disjoint.

5.6. We now define the predicates T_i. We deviate from the outline of the proof presented in §5.2, in that $T_i(f)$ will be defined for f a map of $\aleph_1^L$ into itself.

So let $f: \aleph_1^L \to \aleph_1^L$. Then $T_i(f)$ expresses the following:

$$(\forall\alpha)\quad (f \text{ is almost disjoint from } S(f_{\alpha,\beta}) \text{ iff } \beta = i).$$

Let i be a *positive* integer. We define, in M, a set of conditions, Q_i, which "adds an f such that $T_i(f)$".

A typical element of Q_i is an ordered pair $\langle q_0, q_1 \rangle$. Here q_0 is a map with domain a countable ordinal and range included in $\aleph_1^L$. The set q_1 is a *finite* collection of pairs of the form $\langle f_{\alpha,i}; \gamma \rangle$ where $\alpha < \aleph_2^L, \gamma < \aleph_1^L$. Moreover q_0 and q_1 are required to satisfy the following consistency requirement: if $\langle f_{\alpha,i};\gamma\rangle \in q_1$, $\lambda \in$ domain (q_0) and $\gamma < \lambda$, then $q_0(\lambda) \neq S(f_{\alpha,i})(\lambda)$.

We order Q_i as follows. We put $\langle q_1, q_2\rangle \leqq \langle q_1', q_2'\rangle$ iff $q_1 \subseteq q_1'$ and $q_2 \subseteq q_2'$.

LEMMA. *Let M be our fixed countable transitive model of* ZFC + $V = L$. *Define Q_i, within M, as above. Let G be an M-generic filter on Q_i. Put*

$$g = \bigcup\{q_1 : (\exists q_2)(\langle q_1, q_2\rangle \in G)\}.$$

Then $g\colon \aleph_1^L \to \aleph_1^L$. Moreover

$$M[g] \models T_i(g).$$

We leave the verification of this lemma to the reader.

5.7. We let Q_0 be the set of conditions appropriate to adding a generic bijection, $f\colon \aleph_0 \cong \aleph_1^L$. Thus $f \in Q_0$ iff f is a function whose domain is a finite subset of $\aleph_0$, and f maps its domain one-one into $\aleph_1^L$. We order Q_0 by inclusion. It is easy to verify that if G is an M-generic filter on Q_0, and $g = \cup G$, then g is a bijection of $\aleph_0$ with $\aleph_1^L$.

Let Q be the weak direct product of the Q_i's. Let G be an M-generic filter on Q. By the product lemma, G determines an M-generic filter, G_i, on Q_i. Let $g_0 = \bigcup G_0$. For $i > 0$, let $g_i\colon \aleph_1^L \to \aleph_1^L$ be obtained from G_i as in Lemma 5.6. Finally, let

$$N_1 = M[\langle g_i,\ i \in \omega\rangle].$$

The following lemma is the analog, in the present context, of Theorem 4.8. The proof will be left to the reader. (Cf. the proof of Theorem 4.8.)

LEMMA. *For each $i > 0$, we have $N_1 \models T_i(g_i)$. Let now $j > 0$. Put $g_i' = g_i$, for $i \neq j$. Let $g_j' = \phi$. Let $N^j = M[\langle g_i', i < \omega\rangle]$. Then, $N^j \models (\forall g) \neg T_j(g)$.*

5.8. In $M[g_0]$, $\aleph_1^L$ is countable. Hence there is a set $a_0 \subseteq \omega$ in which every constructible set of integers is recursive.

Next, for each i greater than zero, we define a set $a_i \in M[g_0, g_i]$:

1) $2^r 3^s \in a_i \leftrightarrow g_0(r) < g_0(s)$.
2) $5^r 7^s \in a_i \leftrightarrow g_i(g_0(r)) = g_0(s)$.
3) $n \in a_i$ only as required by 1) and 2).

The reader should verify that $M[a_i] = M[g_0, g_i]$.

We now work inside N_1, constructing a set of integers a from $\{a_i\}$, exactly as in §4.9. We put $N = M[a]$. Clearly, $a_0 \in M[a]$. Moreover, using Lemma 5.6, we see that

$$a = \{j : g_j \in N\} = \{j : (\exists x) T_j(x) \text{ or } j = 0\}.$$

It remains to show that a is a Π_2^1-singleton. This will follow easily, via the techniques used to prove Lemma 4.9.2, from the following lemma.

LEMMA. *There is a predicate* $S(i,x)$, Π_2^1 *as two place predicate, such that*

1) $N \models S(i,a_i)$, *for* $i > 0$.
2) *If* $N \models S(i,x)$, *and* $i > 0$, *then* $i \in a$.

5.8. We turn to the proof of Lemma 5.7. The predicate $S(i,x)$ will be the conjunction of three clauses:

1) i is greater than zero.
2) Let $R = \{\langle r,s\rangle : 2^r 3^s \in a\}$. Then R is a well-ordering of ω of type $\aleph_1^L$.
3) Let $g\colon \omega \cong \aleph_1^L$ be an order isomorphism of $\langle \omega, R\rangle$ with $\aleph_1^L$. Let

$$h = \{\langle g(r), g(s)\rangle : 5^r 7^s \in a\}.$$

Then h is a function, and $T_i(a)$.

From our description of S, everything in Lemma 5.7 is clear except that S is Π_2^1. Clause 1) is clearly Π_2^1.

How about clause 2)? The statement that R is a well-ordering is Π_1^1. Let λ be the order type of R. Then the set of true sentences of L_λ is hyperarithmetic in R. (Cf. [1].) Thus to say that every ordinal $\alpha < \lambda$ is countable in L_λ is a Π_1^1 property of R. This ensures that $\lambda \leqq \aleph_1^L$. Finally to say that $\lambda = \aleph_1^L$, we say that if R' codes a well-ordering of type λ', and every $\alpha < \lambda'$ is countable in $L_{\lambda'}$, then $\lambda' \leqq \lambda$. This statement is easily checked to be Π_2^1.

We turn to clause 3). It is fairly easy to construct a Π_1^1 predicate, $B(R,S)$, which expresses the following:

R and S are binary relations on ω; R is a well-ordering; $\langle \omega; S\rangle$ is a well-founded model of $\boldsymbol{T}$, and the order type of the ordinals of $\langle \omega; S\rangle$ is larger than the order type of R. Hence if x satisfies clause 2), the set of codes of well-founded models of $\boldsymbol{T}$ of length greater than $\aleph_1^L$ is Π_1^1 in x. That clause 3) is Π_2^1 may now be seen by an argument similar to the proof of Lemma 4.5 (7).

6. Some open questions

6.1. As we mentioned in the introduction, Theorem 5.1 establishes the consistency, relative to ZFC, of some consequences of measurable cardinals. However, there are some further consequences, which we now describe, whose consistency might be provable by a clever use of the "almost disjoint sets" technique.

(7) Actually, we should introduce a theory $\boldsymbol{T}'$, and use the corresponding fact about codes for models of $\boldsymbol{T}'$.

If there are measurable cardinals, then there is a certain Π_2^1 predicate, $S(a, b)$, such that:

1) For each $a \subseteq \omega$, there is exactly one $b \subseteq \omega$ such that $S(a, b)$.

2) If x, y, z are reals, x is constructible from y, and $S(x, z)$, then y is recursive in z.

Note that 1) and 2) have the consequence:

3) If x, y are reals, and x is constructible from y, then x is Δ_3^1 in y.

This in turn implies

4) If x is a real, $\aleph_1^{L[x]}$ is countable.

Finally, 4) has the well-known consequence:

5) $\aleph_1$ is inaccessible in $L[x]$ for any real x.

Thus a natural extension of Theorem 1 would be the following:

Suppose that ZFC + "There is an inaccessible cardinal" is consistent. Show that one can find a Π_2^1 predicate, S, such that ZFC + 1) + 2) is consistent. Less ambitiously, show that ZFC + 3) is consistent.

6.2. Our next problem is related to another property of $O^{\#}$. Let α be a cardinal in $L[O^{\#}]$. Then α is inaccessible in L.

We ask if the corresponding consistency results can be obtained without the use of measurable cardinals. For definiteness, we ask the following. Suppose ZFC + "There is a Mahlo cardinal" is consistent. Then is the following theory consistent: ZFC + "Every cardinal is inaccessible in L" + "There is a real, a, from which every set is constructible"?

6.3. Let M be a countable transitive model of ZFC. Let x be a real. We suppose that $M[x]$ is again a model of ZFC and that the cardinals of M are precisely the cardinals of $M[x]$. Suppose finally that GCH holds in M. We ask if GCH holds in $M[x]$.

There are two known cases in which we can conclude that GCH holds in $M[x]$:

1) If $V = L$ holds in M. More generally, if, in M, there is a subset of $\aleph_1$ from which every set is constructible.

2) If $M[x]$ is obtained from M via forcing and the relevant set of conditions is $\aleph_1$-saturated.

It is natural to ask if GCH holds in $M[x]$ without any condition such as 1) or 2). We conjecture that it does not. Possibly the methods of the present paper can produce a counterexample.

6.4. We close the paper with a brief sketch of a result which makes us optimistic about getting counterexamples of the type asked for in §6.3. Let M be a countable standard model of ZFC + $V = L$. Then there is a real a such that in $M[a]$, $\aleph_3^M$ is collapsed down to $\aleph_2^M$ but all other cardinals are preserved. (This result was noticed independently by Silver and Solovay.)

Proof. (Sketch) We collapse $\aleph_3$ down to $\aleph_2$ in the usual way. Call A_1 a subset of $\aleph_2$ that codes the collapsing map. By a variant of the "almost disjoint set" trick using subsets of $\aleph_1$, we construct a subset A_2 of $\aleph_1$ such that $A_1 \in M[A_2]$.

The conditions used to construct A_2 are $\aleph_2$-saturated and $\aleph_1$-closed. Hence $M[A_1]$ and $M[A_2]$ have the same cardinals. Finally, we use the usual "almost disjoint set" trick to produce a subset A_3 of $\aleph_0$ such that $A_2 \in M[A_3]$, and $M[A_2]$ and $M[A_3]$ have the same cardinals. This completes the sketch.

REFERENCES

[1] J. W. Addison, *Some consequences of the axiom of constructibility*, Fund. Math., **46** (1958), 337–357.

[2] K. Gödel, *The consistency of the axiom of choice and of the generalized continuum-hypothesis with the axioms of set theory*, Ann. Math. Studies no. 3, second printing, Princeton 1951.

[3] R. B. Jensen, *Definable sets of minimal degree*, these Proceedings, pp. 122–128.

[4] D. A. Martin and R. M. Solovay, *Internal Cohen Extensions*, to appear.

[5] K. McAloon, Doctoral dissertation, University of California, Berkeley, 1966.

[6] J. R. Shoenfield, *Unramified forcing*, to appear in the Proc. U.C.L.A. Summer Institute on Set Theory.

[7] ———, *The problem of predicativity*, Essays on the foundations of mathematics, 132–139, Jerusalem Academic Press, Jerusalem, 1961.

[8] R. M. Solovay, *A model of set-theory in which every set of reals is Lebesgue measurable*, Ann. of Math., to appear.

[9] ———, *A nonconstructible Δ_3^1 set of integers*, Trans. Amer. Math. Soc. **127** (1967), 50–75.

ON LOCAL ARITHMETICAL FUNCTIONS AND THEIR APPLICATION FOR CONSTRUCTING TYPES OF PEANO'S ARITHMETIC

HAIM GAIFMAN

§0. Introduction.

The aim of this paper is to show how one can use properties of local functions in order to get results concerning models of Peano's arithmetic. The concept of a local relation (or function) is defined in §1. The basic property of these functions is that the recursive definition which defines their iteration can be formalized in Peano's arithmetic, a property which does not hold, in general, for arithmetical functions whose arguments and values are sets of natural numbers. This property and similar ones are proved in §1. In §2 it is shown how the results of §1 can be employed in order to get certain types of elements with respect to the theory of Peano's arithmetic. Special cases of the theorems of §1 were proved, or implicitly used, before, e.g. in [1]. It is, however, interesting to treat the general concept of a local function, especially so since it has natural analogues in other theories, e.g., in number theory with second-order quantifiers and in set theory. In the last case, the concept was indeed used to get results concerning measurable cardinals, the proof of which involved iterating a function whose arguments and values were proper classes, see [3]. The rest of the introduction is devoted to the outlining of some of the results concerning models and types of Peano's arithmetic, which are made possible through the use of local functions. It contains also the definition of minimal types, a concept which occurs later in §3.

MacDowell and Specker have shown in [1] that every model of Peano's arithmetic has an end-extension, where by an "end-extension" a proper elementary extension is meant, in which every new element is bigger in the natural ordering than every old element.

The result and the methods of MacDowell and Specker were improved by the author in several directions. It turns out that one can develop a theory which deals with the structure and the possible extensions of models of Peano's arithmetic. The basic concepts of such a theory are certain types of Peano's arithmetic, where by a type, t, written also "$t(v)$", we mean a maximal set of formulas with one free variable, say v, consistent with

Peano's arithmetic. We consider also types over models of Peano's arithmetic. These are defined in a similar manner, except that the formulas are allowed to have parameters ranging over names of the elements of the model in question, and the type is required to be consistent with the complete diagram of that model. These are well known concepts from the theory of models. Other well known concepts, such as the realization of a type by an element of a model, or the concept of a Skolem function, are assumed to be known. Since one has in Peano's arithmetic definable Skolem functions, there exists, given any model and any set of its elements, the elementary submodel generated by this set. All the models which are mentioned are assumed to be models of Peano's arithmetic and "$<$" denotes their natural ordering. "$v_1 < v_2$" denotes also the formula which asserts that v_1 is smaller than v_2. Similarly, "$v_1 = v_2$" denotes, also, a formula.

A type $t(v)$ is *unbounded* if $\tau < v$ is in t whenever τ is a term of Peano's arithmetic without free variables.

Similarly, a type t over a model is unbounded if the same condition holds, except that now names of the elements of the model are, and may occur in, constant terms.

It can be shown that if an unbounded type of Peano's arithmetic is consistent with the complete diagram of a model, then it can be extended to an unbounded type over that model. Usually it has more than one extension.

A type t is an *end-extension* type if it is unbounded and, for every model $\mathfrak{A}$, if the model $\mathfrak{B}$ is an elementary extension of $\mathfrak{A}$ which is generated by the members of $\mathfrak{A}$ together with one additional element, b, which is bigger than every member of $\mathfrak{A}$ and realizes t, then $\mathfrak{B}$ is an end-extension of $\mathfrak{A}$.

A type t is *minimal* if it is unbounded and, for every model $\mathfrak{A}$, if $\mathfrak{B}$ is related to $\mathfrak{A}$ as in the preceding definition, then $\mathfrak{B}$ is a minimal extension of $\mathfrak{A}$; that is, every model $\mathfrak{C}$ which is an elementary submodel of $\mathfrak{B}$ and includes $\mathfrak{A}$ is either $\mathfrak{B}$ or $\mathfrak{A}$.

A type t is *antiminimal* if it is unbounded and for every model $\mathfrak{A}$, if $\mathfrak{B}$ is related to $\mathfrak{A}$ as in the preceding definition, then $\mathfrak{B}$ is not a minimal extension of $\mathfrak{A}$.

A type $t_1(v)$ *depends on* a type $t_2(v)$ if there is a term, $\tau(v)$, with one free variable such that $t_1(v)$ consists of all the formulas $\phi(v)$ for which $\phi(\tau(v))$ is in $t_2(v)$. For such a τ we put: $t_1 = \tau(t_2)$.

It should be remarked here that we have in our language the minimization operator, namely, if ϕ is any formula then $\mu v \phi$ is a term whose intended meaning is rendered by the reading: the smallest v such that ϕ, if there is such a v, and 0 otherwise.

The import of the notion of dependency becomes clear when it is noted that, if b realizes t in a model $\mathfrak{B}$, then the value in $\mathfrak{B}$ of $\tau(b)$ realizes $\tau(t)$

Similarly, if b realizes in $\mathfrak{B}$ a type t over an elementary submodel, $\mathfrak{A}$, then the value of $\tau(b)$ realizes $\tau(t)$, where τ may have names of members of $\mathfrak{A}$ occurring in it. Thus, the types over $\mathfrak{A}$ which are realized in the model generated by the members of $\mathfrak{A}$ and by b are exactly those which depend over $\mathfrak{A}$ on the type which is realized by b. Note that dependency is reflexive and transitive: If t_1 depends on t_2 and t_2 on t_3, then t_1 depends on t_3.

The following have been proved.

Every end-extension type, t, has exactly one extension to an unbounded type over any model $\mathfrak{A}$ whose complete diagram is consistent with t. Moreover, the property of having unique unbounded extensions characterizes end-extension types: If an unbounded type t is such that, for every model $\mathfrak{A}$ whose complete diagram is consistent with t, there is a unique extension of t to an unbounded type over $\mathfrak{A}$, then t is an end-extension type.

(Note that given any model $\mathfrak{A}$ and any type over $\mathfrak{A}$, there is, up to an isomorphism, exactly one elementary extension of $\mathfrak{A}$ which is generated by the members of $\mathfrak{A}$ and an additional element realizing this type.)

Every minimal type is an end-extension type, but not vice versa, for there are end-extension types which are antiminimal.

If $t_1(v)$ depends on $t_2(v)$ and $t_2(v)$ is an end-extension type then either $t_1(v)$ is discrete, by which we mean that for some constant term, τ, $v = \tau$ is in t_1, or else t_1 is also an end-extension type.

The same is true with "end-extension" replaced everywhere by "minimal". Moreover, in the case that t_1 depends on t_2 and they are minimal types, it is not difficult to deduce from the definition of minimal types that t_2 depends also on t_1. Thus, for minimal type dependency is an equivalence relation.

If t_1 and t_2 are minimal independent types and t_1' and t_2' are unbounded extensions of t_1 and t_2, respectively, over a model $\mathfrak{A}$, then t_1' and t_2' are independent over $\mathfrak{A}$. If $\mathfrak{B}_i$, $i = 1, 2$, is an elementary extension of $\mathfrak{A}$ which is generated by the members of $\mathfrak{A}$ and one additional element which realizes t_i' then every two members from $\mathfrak{B}_1$ and $\mathfrak{B}_2$, respectively, which do not belong to $\mathfrak{A}$, realize different types over $\mathfrak{A}$.

For any type $t(v)$ and any term $\tau(v)$, either $\tau(v) = v$ is in t or else $\tau(t) \neq t$. The same is true for types over a model, in which case names of elements of the model may occur in τ.

Let $X = (X, <)$ be an ordered set and let f be a function assigning to every member, x, of X an end-extension type, $f(x)$, such that $\{f(x) : x \in X\}$ is a set of types of the same complete theory; then, if $\mathfrak{A}$ is any model of that complete theory, whose domain, A, is disjoint from X, there exists an elementary extension, $\mathfrak{B}$, of $\mathfrak{A}$ which is generated by $A \cup X$, in which x realizes $f(x)$, for each $x \in X$, and in which the natural ordering, restricted to X, is $<$. The model $\mathfrak{B}$ is, up to an isomorphism over $\mathfrak{A}$, completely determined by $(X, <)$ and the function f. Moreover, one can specify exactly the automorphisms of $\mathfrak{B}$ over $\mathfrak{A}$, and if, for each x, $f(x)$ is a minimal type, one can specify exactly all the elementary submodels of $\mathfrak{B}$ which include $\mathfrak{A}$, namely: every elementary submodel of $\mathfrak{B}$ which includes $\mathfrak{A}$ is generated by a set of the form $A \cup X'$, where $X' \subset X$, and different subsets of X give rise to different models. One gets also information on the order type of $\mathfrak{B}$, cf. [2].

These results, and others of the same kind, are of interest provided that the types mentioned exist. Now, concerning existence, it has been proved that there are $2^{\aleph_0}$ minimal types and $2^{\aleph_0}$ antiminimal types; moreover, the set of all minimal (antiminimal) types is dense in the set of all bounded types, that is, if $\phi(v)$ is any formula with one free variable such that $\phi(v)$ is consistent with the set of all formulas of the form $\tau < v$, where τ is a constant term, then there is a minimal (antiminimal) type containing ϕ as a member. Actually, there are $2^{\aleph_0}$ such types containing ϕ. Concerning independence of types, it can be shown that $2^{\aleph_0}$ end-extension types exist, which are mutually independent and of which $2^{\aleph_0}$ are minimal and $2^{\aleph_0}$ are antiminimal.

All the results are valid for any theory which extends Peano's arithmetic. The language of the theory may have countably many relation and operation symbols, provided that the induction schema ranges over all the formulas of the language. There is no restriction on the cardinality of the set of individual constants and the more individual constants we have the more end-extension types one can construct. The result is as follows: If A is a set of constant terms such that, for any two different terms, τ_1 and τ_2, of A, the sentence $\neg(\tau_1 = \tau_2)$ is a theorem in the theory in question, then there are $|A|^{\aleph_0}$ mutually independent minimal (antiminimal) types. (It can be shown that there are at most $|C|^{\aleph_0}$ end-extension types, where C is the set of all constant terms, hence the preceding result cannot be improved.)

All the results concerning the existence of types, or of families of types, rely for their proof on the possibility of formalizing certain constructions in local functions which are described in §1.

§1. Local functions and their iterations.

$i, j, k, l, m, n, w, x, y, z$ range over natural numbers; X, Y over sets of natural numbers; u, v are individual variables and U, V are predicate variables of the formal language.

Let Φ be an operation such that, for every set, X, of natural numbers, $\Phi(X)$ is a set of natural numbers. The nth iteration of Φ, Φ^n, is determined by the following recursion:

$$\Phi^0(X) = X,$$

$$\Phi^{n+1}(X) = \Phi(\Phi^n(X)).$$

Assume that Φ is arithmetical, that is, the relation $\lambda x X(x \in \Phi(X))$ can be defined by a formula of the language of Peano's arithmetic in which x and X are represented by an individual variable and a predicate variable, respectively. (We have in our language predicate variables so that, whenever U is a predicate variable and u an individual variable or constant, $U(u)$, written also as "$u \in U$", is a formula. But we do not have second-order quantifiers.)

For every fixed n, Φ^n is arithmetical as well. On the other hand, the relation $\lambda xyX(x \in \Phi^y(X))$ is, in general, not arithmetical. It is not difficult to construct an arithmetical Φ such that, for every X, $\Phi(X) \subset X$, and such that the set of all sentences which are true in the standard model of natural numbers is arithmetical in $\lambda xy(x \in \Phi^y(\omega))$, where ω is the set of all natural numbers. Consequently, in this case, $\lambda xy(x \in \Phi^y(\omega))$ and, a fortiori, $\lambda xyX(x \in \Phi^y(X))$ are not arithmetical.

Our aim is to find a not too restrictive condition on Φ which will imply that the relation $\lambda xyX(x \in \Phi^y(X))$ is arithmetical. Moreover, we also want to have an analogous condition on the formula which is supposed to define, in the language of Peano's arithmetic, the relation $\lambda xX(x \in \Phi(X))$, so that a formula which is supposed to define $\lambda xyX(x \in \Phi\ (X)$ exists, such that the recursive definition of Φ^n, when expressed with the help of these formulas, is a theorem of Peano's arithmetic (or the particular extension of Peano's arithmetic with which we deal).

From now on, ω is the set of natural numbers. $\{0,1\}^\omega$, written also as "2^ω", is the set of all subsets of ω; these are conceived as characteristic functions so that, for $X \in 2^\omega$, "$X(i) = 0$" is equivalent to "$i \in X$" and "$X(i) = 1$" is equivalent to "$i \notin X$." We define $X \restriction x$ to be the sequence $\langle X(0), \ldots, X(x-1) \rangle$, if $x > 0$, and the empty sequence, if $x = 0$.

Definition. Let Q be a relation, such that $Q \subset \omega \times 2^\omega$, then Q is *local* if there are two arithmetical relations on $\omega \times \omega$, say R_1 and R_2, such that, for all X and x, we have

(1) $$Q(x, X) \Leftrightarrow \exists y, m[y = X \restriction m \,\&\, R_1(x, y)],$$

(2) $$\neg\, Q(x, X) \Leftrightarrow \exists y, m\,[y = X \restriction m \,\&\, R_2(x, y)]$$

("$\neg$", "$\forall$" and "$\exists$" are used also in the metalanguage).

Here, "$y = X \restriction m$" means that the number y represents the finite sequence $X \restriction m$. (Using one of the usual encoding methods, we encode the finite sequences of natural numbers and regard the natural numbers as these sequences.)

Roughly speaking, the property of being local means that, in order to decide whether or not $Q(x, X)$, only a finite amount of information on X is needed. By checking certain arithmetical conditions for the finite sequences $\langle X(0), \ldots, X(m-1)\rangle$, one is bound to get, eventually, either a positive or a negative answer*. Obviously, every local relation is arithmetical, but not vice versa. Every recursive relation is local but not vice versa.

DEFINITION. A function $\Phi: 2^\omega \to 2^\omega$ is *local* if $\lambda x X[x \in \Phi(X)]$ is a local relation.

From now on, $\boldsymbol{P}$ is any extension of Peano's arithmetic.

DEFINITION. A formula $\phi(u, U)$ in which u and U are a free individual and a predicate variable, respectively, is said to be ***P**-provably local* if two formulas $\rho_1(u, v)$ and $\rho_2(u, v)$ exist, with u and v as the only free variables, such that the formulas expressing the equivalences (1) and (2) are theorems of $\boldsymbol{P}$. In these formulas, ϕ is supposed to define $\lambda x X Q(x, X)$, with u corresponding to x and U to X, and ρ_i $(i = 1, 2)$ defines R_i. (Thus, the formula asserting (1) would be $\phi(u, U) \leftrightarrow \exists v, v'[v = U \restriction v' \wedge \rho_1(u, v)]$.)

Instead of "***P**-provably local*", we will write, for simplicity, "provably local".

THEOREM 1. (I) *If Φ is local then $\lambda x y X(x \in \Phi^y(X))$ is arithmetical.*

(II) *If $\phi(u, U)$ is provably local then, regarding ϕ as asserting that $x \in \Phi(X)$, where u corresponds to x and U to X, the definition of Φ^n can be formalized in $\boldsymbol{P}$, in the sense that a formula $\phi^*(u, v, U)$ exists such that the recursive conditions for Φ^n formulated with the help of ϕ and ϕ^* are theorems of $\boldsymbol{P}$. (Here $\phi^*(u, v, U)$ is to be regarded as asserting that $x \in \Phi^y(X)$, with u corresponding to x, v to y and U to X.)*

Proof. (I) Consider finite sequences of 0's and 1's. If z is such a sequence we write "$z = \langle z(0), \ldots, z(x-1)\rangle$", where x is the length of z and, for $i < x$, $z(i)$ is the ith member of z. The length of z will be denoted by "$lh(z)$" and we assume that 0 is the sequence of length 0, i.e., the empty sequence. Define $z \restriction y$ to be $\langle z(0), \ldots, z(x-1)\rangle$, where $x = \min(y, lh(z))$. If, for some y, we have $z_1 = z_2 \restriction y$, then we put: $z_1 \prec z_2$, i.e. z_2 extends z_1.

Let R_1 and R_2 satisfy (1) and (2). Say that a sequence z decides x positively if there exists an m such that $m \leq lh(z)$ and $R_1(x, z \restriction m)$. Similarly, z decides x negatively if there exists $m \leq lh(z)$ and $R_2(x, z \restriction m)$. Say that z decides x if it either decides x positively or negatively.

* This concept can be generalized by requiring R_i $(i = 1, 2)$ to be analytic. In set theory one gets an analogous definition by regarding X as a proper class, m, x, y as sets, $X \restriction m$ as $X \cap m$ and R_i as properties expressible in first-order set theory.

From the definitions it follows that if z is a sequence of 0's and 1's and z decides x positively (negatively), then so does every sequence which extends z. From (1) and (2) it follows that no sequence of 0's and 1's decides x both positively and negatively and for every X there is an m such that $X \upharpoonright m$ decides x.

(From the fan theorem it follows that for every x there exists an n such that every sequence of length n decides x, but we will not need this observation.)

Given a sequence, z, of 0's and 1's, let n be the largest number such that $n \leqq lh(z)$ and, for every i, if $i < n$ then z decides i. Let $\Psi(z)$ be defined as the sequence whose length is n, such that, for all $i < n$, $\Psi(z)(i) = 0$ if z decides i positively and $\Psi(z)(i) = 1$ if z decides i negatively. For numbers z which are not sequences of 0's and 1's put: $\Psi(z) = 0$. The following takes place:

(3) $z_1 \precsim z_2 \Rightarrow \Psi(z_1) \precsim \Psi(z_2)$, where z_1 and z_2 are sequences of 0's and 1's,

(4) $\Psi(X \upharpoonright m) \precsim \Phi(X) \upharpoonright m$,

(5) for every X and n, there exists an m such that the length of $\Psi(X \upharpoonright m)$ is at least n.

(3) follows from the definition and the properties of Ψ; (4) follows from (1) and (2). To prove (5), let m_i be such that $X \upharpoonright m_i$ decides i; given n, let m be greater than all m_i, $i < n$.

(3), (4) and (5) imply that $\Psi(X \upharpoonright m)$, as $m = 0, 1, \ldots$ are increasing sequences of 0's and 1's, whose union is the sequence corresponding to $\Phi(X)$.

Since $\lambda z \Psi(z)$ is an arithmetic function from ω into ω, its nth iterate, as defined by the usual recursion, can be given an explicit arithmetic definition. Hence, $\lambda z n \Psi^n(z)$ is arithmetic.

By induction on k it follows from (3) that, for every k and every two sequences, z_1 and z_2, of 0's and 1's, we have:

(6) $$z_1 \precsim z_2 \Rightarrow \Psi^k(z_1) \precsim \Psi^k(z_2).$$

We have also:

(7) For every X and n there exists an m such that the length of $\Psi^k(X \upharpoonright m)$ is at least n.

(7) is proved by induction on k; assume it to be true for k and let Y be the set such that the sequence $Y(0), \ldots, Y(n), \ldots$ is the union of the sequences $\Psi^k(X \upharpoonright m)$, $m = 0, 1, \ldots$, i.e., $x \in Y \Leftrightarrow \exists m[\Psi^k(X \upharpoonright m)(x) = 0]$. Apply (5) to the case $X = Y$ and let m be such that $\Psi(Y \upharpoonright m)$ has a length which is at least n. By the induction hypothesis for k there is an m^* such that $Y \upharpoonright m \precsim \Psi^k(X \upharpoonright m^*)$. Hence $\Psi^{k+1}(X \upharpoonright m^*)$ has a length which is at least n.

Let $\Phi^*(n,X)$ be the set such that $\Phi^*(n,X)(0),\ldots,\Phi^*(n,X)(i),\ldots$ is the union of the sequences $\Psi^n(X \upharpoonright m)$, $m = 0,1,\ldots$, i.e.,

$$(8) \qquad x \in \Phi^*(n,X) \Leftrightarrow \exists m[\Psi^k(X \upharpoonright m)(x) = 0].$$

We now prove that $\Phi^*(n,X)$ satisfies the recursion which defines $\Phi^n(X)$. For $n = 0$ we have $\Phi^*(0,X) = X$. By (3), (4) and (5) it follows that if $m_1 \leqq m_2$ then $\Psi(\Phi^*(n,X) \upharpoonright m_1) \prec \Psi(\Phi^*(n,X) \upharpoonright m_2)$ and that the union of the sequences $\Psi(\Phi^*(n,X) \upharpoonright m)$, $m = 0,1,\ldots$ is the sequence which corresponds to $\Phi(\Phi^*(\prec,X))$. On the other hand, for every m there exists an m^* such that $\Phi^*(n,X) \upharpoonright m \prec \Psi^n(X \upharpoonright m^*)$. Hence, the union of the sequences $\Psi(\Phi^*(n,X) \upharpoonright m)$, $m = 0,1,\ldots$, is also the union of $\Psi^{n+1}(X \upharpoonright m)$, $m = 0,1,\ldots$. This proves that

$$(9) \qquad \Phi(\Phi^*(n,X)) = \Phi^*(n+1,X).$$

Hence, $\Phi^n(X) = \Phi^*(n,X)$ and $\Phi^*(n,X)$ is arithmetic, which proves (I).
(II) If $\phi(u,U)$ is provably local then we have formulas $\rho_i(u,v)$ $(i = 1,2)$ such that the equivalences (1) and (2), when formulated using ϕ, ρ_1 and ρ_2, are theorems of $\boldsymbol{P}$. Now, the whole proof of (I) can be formulated within $\boldsymbol{P}$. We get a formula which corresponds to $\lambda xz[x \in \Psi(z)]$ and from it we get a formula which corresponds to $\lambda xyz[x \in \Psi^y(z)]$. Finally, using the formal version of the right side of (8) we get $\phi^*(u,v,U)$ which corresponds to $\lambda xyX[x \in \Phi^*(y,X)]$. The arguments which were just given can be formalized within $\boldsymbol{P}$ so as to a yield a proof in $\boldsymbol{P}$ of the formula which asserts that the induction schema for Φ^n holds.

Q.e.d.

Instead of relations which are subsets of $\omega \times 2^\omega$, consider now any subsets of $\omega^k \times (2^\omega)^l$. The definition generalizes as follows.

DEFINITION. Assume that $Q \subset \omega^k \times (2^\omega)^l$, then Q is local in the jth set-coordinate $(1 \leqq j \leqq l)$, if there are arithmetical relations $R_i (i = 1,2)$ such that $R_i \subset \omega^m \times (2^\omega)^{j-1} \times \omega \times (2^\omega)^{l-j}$ $(i = 1,2)$ and for all $x_1,\ldots,x_k$ and all $X_1,\ldots,X_l$ we have:

$$Q(\vec{x},\vec{X}) \Leftrightarrow \exists z,m[z = X_j \upharpoonright m \wedge R_1(\vec{x},\vec{X}',z,\vec{X}'')],$$

$$\neg\, Q(\vec{x},\vec{X}) \rightarrow \exists z,m[z = X_j \upharpoonright m \wedge R_2(\vec{x},\vec{X}',z,\vec{X}'')].$$

Here, '$\vec{x}$' and '$\vec{X}$' stand for '$x_1,\ldots,x_k$' and '$X_1,\ldots,X_l$', respectively, '$\vec{X}'$' stands for '$X_1,\ldots,X_{j-1}'$' and '$\vec{X}''$' for '$X_{j+1},\ldots,X_l$'.

We will say that $Q(x_1,\ldots,x_k,X_1,\ldots,X_j,\ldots,X_l)$ is local in X_j to express the fact that Q is local in the jth coordinate.

DEFINITION. The function $\Phi(x_1, \ldots, x_k, X_1, \ldots, X_l)$ *is local in* X_j if $\lambda\vec{x}\vec{X}\,[x \in \Phi(x_1, \ldots, x_k, X_1, \ldots, X_l)]$ is local in X_j.

It is obvious how to define the property that a formula $\phi(u_1, \ldots, u_k, U_1, \ldots, U_l)$.is *provably local in* U_j.

THEOREM 2. (I) *If* $\Phi(X)$ *is local in* X *then the iteration* $\Phi^n(X)$, *as a function of* n *and* X, *is also local in* X.

(II) *If* $\phi(u, U)$ *is provably local then the formula* $\phi^*(u, v, U)$ *of Theorem* 1, *which is supposed to define the iterate of the function defined by* ϕ, *is provably local in* U.

Proof. (I) From the proof of Theorem 1 it follows that

$$x \in \Phi^n(X) \Leftrightarrow \exists z, m\,[z = X \upharpoonright m \;\&\; \Psi^n(z)(x) = 0],$$

$$x \notin \Phi^n(X) \Leftrightarrow \exists z, m\,[z = X \upharpoonright m \;\&\; \Psi^n(z)(x) = 1],$$

implying that $\Phi^n(X)$ is local in X.

(II) The formalization of the last two equivalences can be proved in ***P***.

Q.e.d.

Given a function $\lambda xXY\,\Phi(x, X, Y)$, one can form iterates in several ways; we get the following generalization of Theorem 1.

THEOREM 1*. (I) *Assume that* $\Phi(x, X, Y)$ *is local in* X. *Consider* w *as a finite sequence of natural numbers and define* $\Phi^{(w)}(X, Y)$ *by the following recursion*:

$$\Phi^{(0)}(X, Y) = X \quad (\textit{where } 0 = \textit{the empty sequence}),$$

$$\Phi^{(w \frown \langle y \rangle)}(X, Y) = \Phi(y, \Phi^{(w)}(X, Y), Y)$$

(*where* $w \frown \langle y \rangle$ *is the sequence obtained from* w *by concatenation with the sequence* $\langle y \rangle$, *where* $\langle y \rangle$ *is of length* 1 *and its member is* y).

Under these assumptions, $\Phi^{(w)}(X, Y)$, *as a function of* w, X *and* Y. *is arithmetic and it is local in* X.

(II) *If* $\phi(u, v, U, V)$ *is provably local in* U, *then, regarding* ϕ *as asserting that* $x \in \Phi(y, x, Y)$, *the definition of* $\Phi^{(w)}(X, Y)$ *can be formalized in* ***P***, *in the sense that a formula* $\phi^*(u, v, U, V)$ *exists such that the formula corresponding to the recursion schema is provable in* ***P***. *The formula* $\phi^*(u, v, U, V)$ *is provably local in* U.

The proof of Theorem 1* uses the same construction as the proof of Theorem 1. The difference is that we now have additional parameters. Thus, instead of defining "z decides x" one defines "z, y and Y decide x" (where z is a sequence of 0's and 1's). The function Ψ will now be a function of three arguments, the sequence, z, of 0's and 1's and the two additional parameters y and Y. Instead of $\Psi^n(z)$ we now have $\Psi^{(w)}(z, Y)$, defined by:

$$\Psi^{(0)}(z, Y) = z,$$

$$\Psi^{(w\frown\langle y\rangle)}(z, Y) = \Psi(y, \Psi^{(w)}(z, Y), Y).$$

The function $\Phi^{(w)}(X, Y)$ can be defined as $\Phi^*(w, X, Y)$, where $\Phi^*(w, X, Y)$ is such that the sequence $\Phi^*(w, X, Y)(0)$, $\Phi^*(w, X, Y)(1), \ldots$ is the union of the sequences $\Psi^{(w)}(X \upharpoonright m, Y)$, $m = 0, 1, \ldots$. One shows, exactly in the same way as in the proof of Theorem 1, that $\Phi^*(w, X\, Y)$ satisfies the required recursion. The proof of (II) is analogous.

It is obvious that we may assume that Φ has also additional arguments ranging over ω or 2^ω. In this case, the arguments will be also arguments of $\Phi^{(w)}(X, Y)$; the proof is not affected.

THEOREM 3. (I) *Let $\Phi(x, X)$ be a function which is local in X, and let g be an arithmetic function such that, for every x and X, $g(x, X)$ is a number. Define $\Phi_g^n(X)$ by the recursion:*

$$\Phi_g^0(X) = X,$$

$$\Phi_g^{n+1}(X) = \Phi(g(n, \Phi_g^n(X)), \Phi_g^n(X)).$$

Then $\lambda n X \Phi_g^n(X)$ is arithmetic.

(II) *Let $\phi(u, v, U)$ be provably local in U and let $\gamma(u, U)$ be a term. Then, regarding ϕ as defining the relation $\lambda x X[x \in \Phi(y, X)]$ and γ as defining g, there exists a formula $\phi_\gamma^*(u, v, U)$ such that the recursive schema above, when formalized using ϕ, γ and ϕ_γ^*, is provable in* $\boldsymbol{P}$. *(In this formalization $\phi_\gamma(u, v, U)$ should correspond to $x \in \Phi_g^n(X)$ with u, v and U corresponding to x, n and X, respectively.)*

It should be noted that Φ, as well as g, might have other arguments ranging over ω or 2^ω. These are to be treated as parameters and they will occur also in $\Phi_g^n(X)$. Similarly, ϕ and γ might have other free variables, in which case they will appear in ϕ_γ^*.

Proof. (I) By Theorem 1*, the function $\lambda w X \Phi^{(w)}(X)$ which is defined by the schema

$$\Phi^{(0)}(X) = X,$$

$$\Phi^{(w\frown\langle y\rangle)}(X) = \Phi(y, \Phi^{(w)}(X))$$

is arithmetic.

Define a function $h\colon \omega \times 2^\omega \to \omega$ by ordinary course-of-values induction as follows:

$$h(0, X) = g(0, X),$$

$$h(n+1, X) = g(n+1, \Phi^{(\bar{g}(n+1, X))}(X)),$$

where $\bar{g}(k, X) = \langle g(0, X), \ldots, g(k-1, X)\rangle$.

Since the induction here is on the numerical argument, the argument X appearing only as a fixed parameter, it follows that h is arithmetical.

It is easy to prove that $\Phi^{(\bar{h}(n,X))}(X)$, as a function of n and X, satisfies the recursion which defines $\Phi_g^n(X)$. Hence, $\Phi_g^n(X) = \Phi^{(\bar{h}(n,X))}(X)$, implying that $\Phi_g^n(X)$ is arithmetical.

(II) The construction carried out in (I) can be formalized in $\boldsymbol{P}$, because h is defined by ordinary course-of-values recursion. Thus, we get a term which corresponds to h, from which, together with the formula which corresponds to $\Phi^{(w)}(X)$, we get the formula corresponding to $\Phi^{(\bar{h}(n,X))}(X)$. The proof that the recursion holds can then be formalized in $\boldsymbol{P}$. Q.e.d.

Theorem 3 can be generalized by adding to the function g an additional argument ranging over 2^ω so that instead of $g(x, X)$ we have $g(x, X_1, X_2)$, and taking as the recursive definition of $\Phi_g^n(X)$, where $n \geqq 1$, the following formula:

$$\Phi_g^{n+1}(X) = \Phi(g(n, \Phi_g^n(X), \Phi_g^{n-1}(X)), \Phi_g^n(X)),$$

while the case $n = 0$ is taken care of by defining $\Phi_g'(X)$ to be some given arithmetical function of X. The proof is analogous.

In a similar way we can use $g(x, X_1, \ldots, X_k)$ with an analogous recursive condition for $n \geqq k - 1$. All these cases are special cases of the following recursion:

$$\Phi_g^{n+1}(X) = \Phi(g(\langle \Phi_g^0(X), \ldots, \Phi_g^n(X) \rangle), \Phi_g^n(X)),$$

where $\langle \Phi_g^0(X), \ldots, \Phi_g^n(X) \rangle$ is the set which encodes in it the sets $\Phi_g(X)$ $(i = 0, \ldots, n)$ so that from this set one can recover in a recursive way the number n and the sets $\Phi_g(X)$ $(i \leqq n)$. (One can define, for example, $\langle X_0, \ldots, X_{n-1} \rangle$ to be $\{\langle n, i, j \rangle : i < n \,\&\, j \in X_i\}$.) The theorem is still true for this case and the proof is analogous.

The definition of a relation which is local in a given coordinate can be generalized to that of a relation which is simultaneously local in several coordinates:

Definition. Let Q be a relation such that $Q \subset \omega^k \times (2^\omega)^l$, then $Q(x_1, \ldots, x_k, X_1, \ldots, X_j, Y_1, \ldots, Y_{l-j})$ is said to be *local in* $(X_1, \ldots, X_j)$ if there exist two arithmetic relations R_i, $R_i \subset \omega^k \times \omega^j \times (2^\omega)^{l-j}$ $(i = 1, 2)$, such that the following equivalences hold for all $x_1, \ldots, x_k$ and all $X_1, \ldots, X_j$, $Y_1, \ldots, Y_{l-j}$:

$$Q(\vec{x}, \vec{X}, \vec{Y}) \Leftrightarrow \exists \vec{z}, \vec{m}\,[\vec{z} = \vec{X} \restriction \vec{m} \,\&\, R_1(\vec{x}, \vec{z}, \vec{Y})],$$

$$\neg\, Q(\vec{x}, \vec{X}, \vec{Y}) \rightarrow \exists \vec{z}, \vec{m}\,[\vec{z} = \vec{X} \restriction \vec{m} \,\&\, R_2(\vec{x}, \vec{z}, \vec{Y})].$$

Here, '$\vec{x}$', '$\vec{X}$' and '$\vec{Y}$' stand, respectively, for '$x_1, \ldots, x_k$' '$X_1, \ldots, X_j$' and '$Y_1, \ldots, Y_{l-j}$'; '$\vec{z}$' and '$\vec{m}$' stand, respectively, for '$z_1, \ldots, z_j$' and '$m_1, \ldots, m_j$', and '$\vec{z} = \vec{X} \upharpoonright \vec{m}$' for the conjunction of '$z_i = X_i \upharpoonright m_i$' $(i = 1, \ldots, j)$.

It is easily seen that $Q(\vec{x}, X_1, \ldots, X_j, \vec{Y})$ is local in $(X_1, \ldots, X_j)$ if and only if $Q^*(\vec{x}, X, \vec{Y})$ is local in X, where Q^* is obtained from Q through the encoding of $X_1, \ldots, X_j$ into a single set X.

It is obvious how to define the analogous property of formulas: to be *provably local in* $U_1, \ldots, U_j$.

DEFINITION. The function $\Phi(x_1, \ldots, x_k, X_1, \ldots, X_j, Y_1, \ldots, Y_{l-j})$ *is local in* $(X_1, \ldots, X_j)$ if the relation $\lambda z\vec{x}\vec{X}\vec{Y}[z \in \Phi(\vec{x}, \vec{X}, \vec{Y})]$ is local in $X_1, \ldots, X_j$.

One can form iterations of functions in many variables in several ways. Theorems analogous to Theorems 1, 2, 1* and 3 hold for all these cases. For instance, if $\Phi_1(X, Y)$ and $\Phi_2(X, Y)$ are local in (X, Y) and if Φ_1^n and Φ_2^n are defined by:

$$\Phi_1^{n+1}(X, Y) = \Phi_1(\Phi_1^n(X, Y), \Phi_2^n(X, Y)),$$

$$\Phi_2^{n+1}(X, Y) = \Phi_2(\Phi_1^n(X, Y), \Phi_2^n(X, Y)),$$

then $\Phi_i^n(X, Y)$ as a function of n, X and Y is arithmetic, moreover it is local in (X, Y). Analogous versions of Theorem 3 hold as well. Actually, the cases of several functions which are local in several variables can be reduced to the cases of one function which is local in one variable, by using encoding of finite sequences of sets into single sets.

§2. Construction of minimal types.

The results of §1 are now applied to construct minimal types. Consider an infinite set, X, of natural numbers and a function $f\colon \omega \to \omega$. Either there exists an infinite subset, X', of X, such that the restriction of f to X' is a constant function, or there exists an infinite subset, X', of X, such that the restriction of f to X' is one-to-one. To prove this, we argue as follows. If the first alternative does not hold then, for every w, there are only *finitely* many x's in X such that $f(x) = w$. Hence, for every y there exists a z in X such that, for all x in X, if $x \leqq y$ then $f(x) \neq f(z)$. If z_0 is the first z having this property, then $z_0 > y$ and $f(x) \neq f(z_0)$ whenever $x \in X$ and $x < z_0$. Let $X' = \{z\colon \forall y \in X\ y < z \Rightarrow f(y) \neq f(z)\}$, then it follows that X' is infinite. Obviously, the restriction of f to X' is one-to-one.

The argument just given can be formalized in $\boldsymbol{P}$, where, as before, $\boldsymbol{P}$ any extension of Peano's arithmetic.

Assume now that f is a function of two arguments, $f\colon \omega \times \omega \to \omega$, and X is, as before, an infinite set. Our aim is to construct a sequence of sets $X_0, X_1, \ldots$ as follows. If there exists a number, y, such that, for an infinite subset, X', of X the restriction of $\lambda x f(0,x)$ to X' has the constant value w, let w_0 be the first one and let $X_0 = \{x : x \in X \;\&\; f(0,x) = w_0\}$; otherwise, let X_0 be the infinite subset, X', of X which is constructed as before, such that the restriction of $\lambda x f(0,x)$ to X' is one-to-one. The passage from X_n to X_{n+1} is similar, X_{n+1} is the infinite subset of X_n consisting of all x for which $f(n+1,x) = w_{n+1}$, where w_{n+1} is the first w for which this set is infinite, provided that such a w exists, and if there is no such w then X_{n+1} is the infinite subset of X_n which is constructed as before, such that the restriction of $\lambda x f(n+1,x)$ to X_{n+1} is one-to-one.

Put: $F(n,X) = X_n$. Our aim is to show that if f is arithmetical so is F and, moreover, that the construction can be formalized in $\boldsymbol{P}$ in the sense that, given a term, $\tau(u,v)$, which corresponds to $f(x,y)$, there exists a formula $\phi_\tau(u,v,U)$, corresponding to the relation $\lambda xyX[x \in F(y,X)]$, such that the recursive conditions are provable in $\boldsymbol{P}$.

Let $(\;)_0\colon x \to (x)_0$ and $(\;)_1\colon x \to (x)_1$ be recursive functions from ω onto ω such that, for every pair (y,z), there exists an x for which $(x)_0 = y$ and $(x)_1 = z$. Define the function $\Phi(x,X)$ as a function whose values are subsets of ω which are determined as follows:

If $(x)_1 > 0$ then $z \in \Phi(x,X) \Leftrightarrow z \in X \;\&\; f((x)_0, z) = (x)_1 - 1$.

If $(x)_1 = 0$ then $z \in \Phi(x,X) \Leftrightarrow z \in X \;\&\; \forall y \in X\; y < z \Rightarrow f((x)_0, y) \neq f((x)_1, z)$.

It is easily seen that $X_{n+1} = \Phi(x, X_n)$, where x is such that $(x)_0 = n+1$ and $(x)_1 = w+1$, where w is the smallest number such that $\{z : z \in X \;\&\; f(n+1,z) = w\}$ is infinite, provided that such a number exists, and $(x)_1 = 0$, otherwise. X_0 is obtained from X in a similar way, with 'X' replacing 'X_n' and 'X_0' replacing 'X_{n+1}'.

It is easy to see that if f is arithmetic then $\Phi(x,X)$ is local in X.

Define a function g as follows. $g(n,X)$ is the smallest x satisfying the condition: $(x)_0 = n$ and $(x)_1$ is such that if there exists a y for which $\{z : z \in X \;\&\; f(n,z) = y\}$ is infinite then $(x)_1 = w+1$, where w is the smallest y having that property, and if there is no such y then $(x)_1 = 0$.

If f is arithmetic, so is g. It is obvious that we have:

$$X_0 = \Phi(g(0,X),X),$$

$$X_{n+1} = \Phi(g(n+1,X_n),X_n).$$

Consequently, it follows that, using the notation of Theorem 3, we have:

$$F(n,X) = \Phi_g^{n+1}(X).$$

From Theorem 3(I) it follows that F is arithmetic. The claim concerning the formalization of F within $\boldsymbol{P}$ follows from Theorem 3(II) and the fact that the definition of Φ and the argument proving that it is local can be formalized in $\boldsymbol{P}$.

Now, define $\tilde{f}(n, X)$ and $f^*(n, z, X)$ as follows:

$\tilde{f}(n, \tilde{f}X) = f(n, x_0)$, where x_0 is the smallest member of $F(n, X)$ (this also means that if $F(n, X)$ is empty then $x_0 = 0$).

$f^*(n, z, X)$ is the smallest x in $F(n, X)$ such that $f(n, x) = z$ (again, with the same stipulation concerning the meaning of "smallest x such that...").

If X is infinite then $\lambda x f(n, x)$ restricted to $F(n, X)$ is either constant or one-to-one. Hence, it follows that, if X is infinite, either, for every x in $F(n, X)$, we have $f(n, x) = \tilde{f}(n, X)$, or, for every x in $F(n, X)$, we have $f^*(n, f(n, x), X) = x$.

If f is arithmetic so are $\tilde{f}$ and f^*. It is easy to see that the construction can be formalized so that the formulas expressing the above-mentioned properties of F, $\tilde{f}$ and f^* can be proved in $\boldsymbol{P}$.

For an infinite X the sequence of sets $F(0, x), \ldots, F(n, X), \ldots$ is such that all are included in X, $F(n, X) \supset F(n+1, X)$ and all are infinite. The diagonal of this sequence is the set obtained by taking the 0th member of $F(0, X)$, the 1st member of $F(1, X)$, and so on. Let this set be $D_f(X)$, then:

$$x \in D_f(X) \Leftrightarrow \exists z[x \in F(z, X) \;\&\; \{y \colon y \in F(z, X) \;\&\; y < x\} \text{ has } z \text{ elements}].$$

The formula expressing the right side of the equivalence can be used as the formal definition of $D_f(X)$. It is easy to see that one can prove in $\boldsymbol{P}$ the relevant properties of $D_f(X)$, namely, that if X is infinite then so is $D_f(X)$ and, for every n, $D_f(X)$ is included from a certain point on in $F(n, X)$. This is so because, in the case in which f is given by means of a term of $\boldsymbol{P}$, one can prove in $\boldsymbol{P}$ that, for every n, $F(n, X)$ is infinite if X is, and that $X \supset F(n, X) \supset F(n+1, X)$.

Finally, define $f^{\#}(n, X)$ as the smallest z such that, for all $x > z$, if $x \in D_f(X)$ then $x \in F(n, X)$.

We have:

(10) If X is infinite so is $D_f(X)$.

(11) For every n and every x, if $x \in D_f(X)$ and $x > f^{\#}(n, X)$, then either $f(n, x) = \tilde{f}(n, X)$ or $f^*(n, f(n, x), X) = x$.

Formalizing the construction we get, starting from any term $\tau(u, v)$ which corresponds to f a formula $\Delta(u, U)$ and terms $\tilde{\tau}(u, U)$, $\tau^*(u, v, U)$, $\tau^{\#}(u, U)$, which correspond, respectively, to $\lambda x X[x \in D_f(X)]$ and to $\tilde{f}$, f^* and $f^{\#}$. The following is provable in $\boldsymbol{P}$:

(10*) $[\forall u \exists v > uU(v)] \rightarrow [\forall u \exists v > u\Delta_\tau(v, U)]$,

(11*) $\Delta_\tau(v, U) \wedge v > \tau^{\#}(u, U) \rightarrow [\tau(u,v) = \tilde{\tau}(u, U) \vee \tau^*(u, \tau(u,v), U) = v]$.

Consider now a model, $\mathfrak{A}$, of $\boldsymbol{P}$, whose domain is A, and an elementary extension of it, $\mathfrak{B}$, with the domain B. If $\tau(u_1, \ldots, u_n)$ is a term, let $|\tau(a_1, \ldots, a_n)|_{\mathfrak{B}}$, where $a_1, \ldots, a_n \in B$, be the value of τ in $\mathfrak{B}$, for the substitution of a_i for v_i $(i = 1, \ldots, n)$. If $b \in B$ then the model generated by $A \cup \{b\}$ consists of all the elements of the form $|\tau(a_1, \ldots, a_n, b)|_{\mathfrak{B}}$, where τ ranges over all terms, $n = 0, 1, \ldots$, and $a_1, \ldots, a_n \in A$. Since, using pairing functions, $a_1, \ldots, a_n$ can be encoded into a single element a which will be in A if $a_1, \ldots, a_n$ are, it is enough to let τ range over all terms with two free variables, the first of which is to be substituted by members of A and the second by b.

Now assume that $a < b$ for all $a \in A$. Let $\phi(v)$ be a formula with one free variable such that $\mathfrak{B} \models \forall u \exists v > u\phi(v)$, and let $\tau(u,v)$ be a fixed term. Let $\Delta_{\tau,\phi}(u)$ be obtained from $\Delta_\tau(u, U)$ by substituting everywhere $\phi(v)$ for $U(v)$, where v is any variable. Then, $\mathfrak{B} \models \forall u \exists v > u\,\Delta_{\tau,\phi}(v)$. Let $\tilde{\tau}_\phi(u)$, $\tau^*_\phi(u,v)$ and $\tau^{\#}_\phi(u)$ be obtained in a similar way from $\tilde{\tau}(u, U)$, $\tau^*(u, v, U)$, and $\tau^{\#}(u, U)$. If b is such that $\mathfrak{B} \models \Delta_{\tau,\phi}(b)$, i.e., if b satisfies $\Delta_{\tau,\phi}(v)$ in $\mathfrak{B}$, then from (11*) it follows that, for all $a \in A$, the following is satisfied in $\mathfrak{B}$:

(12) $$b > \tau^{\#}_\phi(a) \rightarrow \tau(a,b) = \tilde{\tau}_\phi(a) \vee \tau^*_\phi(a, \tau(a,b)) = b.$$

If $a \in A$, then $|\tau^{\#}_\phi(a)|_{\mathfrak{B}} \in A$; hence, for $a \in A$, it follows that $b > |\tau^{\#}_\phi(a)|_{\mathfrak{B}}$. Putting $b' = |\tau(a,b)|_{\mathfrak{B}}$ it follows that either $b' = |\tilde{\tau}_\phi(a)|_{\mathfrak{B}}$ or $|\tau^*_\phi(a, b')|_{\mathfrak{B}} = b$. Now, $|\tilde{\tau}_\phi(a)|_{\mathfrak{B}} \in A$. Hence, either b' is in A, or, by applying τ^*_ϕ to a member of A and to b', one can get back b. The second case implies that b belongs to the model generated by $A \cup \{b'\}$.

These considerations show that, if for every term $\tau(u,v)$ there exists a formula $\phi(v)$ such that $\mathfrak{B} \models \Delta_{\tau,\phi}(b)$, then the model generated by $A \cup \{b\}$ is a minimal extension of $\mathfrak{A}$. (Note that if $\mathfrak{B} \models \Delta_{\tau,\phi}(b)$ then, for every $a \in A$, the formula asserting that there exists an $x > a$ satisfying $\Delta_{\tau,\phi}(v)$ holds in $\mathfrak{B}$, consequently it holds in $\mathfrak{A}$. Hence, $\forall u \exists v > u\Delta_{\tau,\phi}(u)$ holds in $\mathfrak{A}$ and, therefore, in $\mathfrak{B}$, implying that $\forall u \exists v > u\phi\,(v)$ holds as well.)

For a type to be minimal it is, therefore, sufficient that it be unbounded and that, for every term $\tau(u,v)$, there exists a formula $\phi(u)$ such that $\Delta_{\tau,\phi}(v)$ is in the type.

Enumerate all terms of the form $\tau(u,v)$, and let the enumeration be:

$$\tau_0, \tau_1, \ldots, \tau_n, \ldots.$$

Let $\phi_0(v)$ be any formula such that $\boldsymbol{P} \vdash \forall u \exists v > u \phi_0(v)$. Define:

$$\Delta_0(v) = \Delta_{\tau_0, \phi_0}(v),$$

$$\Delta_{n+1}(v) = \Delta_{\tau_{n+1}, \Delta n}(v).$$

It follows that for every n, $\boldsymbol{P} \vdash \forall u \exists v > u \Delta_n(v)$ and also $\boldsymbol{P} \vdash \forall v[\Delta_{n+1}(v) \rightarrow \Delta_n(v)]$. Consequently, $\{\Delta_n(v): n = 0, 1, \ldots\} \cup \{v > \sigma: \sigma$ – a constant term$\}$ is consistent with $\boldsymbol{P}$. Every type which includes this set of formulas is minimal.

In order to perform this construction it is sufficient that the language of $\boldsymbol{P}$ should have countably many predicate and function symbols. Individual constants can be disregarded. For, let $\boldsymbol{P}'$ be that part of $\boldsymbol{P}$ which consists of formulas in which no individual constants, besides $\mathbf{0}$ and $\mathbf{1}$, occur. The terms of the language of $\boldsymbol{P}'$ can be enumerated and one can get the set of formulas $\{\Delta_n(u): n = 0, 1, \ldots\}$. This set has also the desired properties with respect to $\boldsymbol{P}$. For, if $\mathfrak{A}$ and $\mathfrak{B}$ are models of $\boldsymbol{P}$, then they are also models of $\boldsymbol{P}'$, the only difference being that as models of $\boldsymbol{P}$ they might have certain elements as distinguished ones. If the model generated by $A \cup \{b\}$ is a minimal extension of $\mathfrak{A}$, when regarded as a model of $\boldsymbol{P}'$, then the same is true when the models are regarded as models of $\boldsymbol{P}$.

Other types, mentioned in the introduction, are shown to exist by formalizing in Peano's arithmetic other set-theoretical constructions. The common feature in all the cases is that the formalization is possible because the operations can be reduced to iterating functions which are local in the sense of §1, so that the theorems of §1, or certain variants of them, apply. In the same way one shows the existence of many types having desired properties. For example, to get a family $\{t^\xi\}_{\xi \in \Xi}$ of mutually independent minimal types, where Ξ is a certain index set, one constructs a doubly-indexed family of formulas $\{\Delta_n^\xi(v): n = 0, 1, \ldots, \xi \in \Xi\}$ such that, for every ξ, the sequence $\Delta_0^\xi(v), \ldots, \Delta_n^\xi(v), \ldots$ behaves in a way which is similar to the behavior of the sequence $\Delta_0, \Delta_1, \ldots, \Delta_n, \ldots$ of this section. To ensure that the types are independent, one has to carry out the construction in such a way that, whenever $\xi \neq \eta$ and whenever $\tau(v)$ is a term with one free variable, there exists an n for which the following is provable in $\boldsymbol{P}$:

$$\exists u \forall v[v > u \wedge \tau(v) > u \rightarrow [\Delta_n^\xi(v) \rightarrow \neg\, \Delta_n^\eta(\tau(u))].$$

The formulas Δ_n^ξ are arrived at by formalizing a process which employs iterations of several functions which are local in several arguments. Variants of Theorem 3, which are mentioned in § 1, are used.

REFERENCES

[1] R. MacDowell and E. Specker, Modelle der Arithmetik, *Infinitistic Methods, Proc. Symp. on Foundations of Mathematics* (Warsaw, 1959), Pergamon Press, London, 1961 pp. 257–263.

[2] H. Gaifman, Uniform extension operators for models and their applications *Sets, Models and Recussion Theory*, Proceedings of the summer school in mathematical logic and tenth logic colloquium (Leicester, Aug.–Sep., 1965), edited by J. Crossley, North-Holland Publ. Co., 1967, pp. 122–155.

[3] H. Gaifman, Pushing up the measurable cardinal. Lecture notes prepared in connection with the summer institute on Axiomatic Set Theory, held at U.C.L.A., Los Angeles, Calif., 1967, Amer. Math. Soc., pp. IV. R 1–IV. R 16.

DEFINABLE SETS OF MINIMAL DEGREE

RONALD JENSEN
Bonn

§1. After Cohen showed that **ZF** is compatible with the existence of a non-constructible real, the question arose whether there can be "tame" non-constructible reals, less drastically removed from Cohen's example. Two obvious measures of tameness are: degree of constructibility and definability. That is, we can ask whether, relative to **ZF**, it is consistent to assume the existence of reals which, though not constructible, have a low degree of constructibility, or of non-constructible reals which are definable in some nice way. Sacks settled the first question by showing that **ZF** is compatible with the existence of a non-constructible real a of minimal degree (i.e. if $x \in L[a]$ and $a \notin L[x]$, then $x \in L$). The earliest answer to the second question was provided by McAloon, who showed that there can be ordinally definable non-constructible reals. After this, attention focused on the question whether non-constructible analytic reals can exist and, if so, at what level of the analytic hierarchy. By Shoenfield's theorem the lowest possible level would be Δ_3^1. Martin and Kripke produced a **ZF** model containing a Δ_4^1 non-constructible real (which was, moreover, of minimal degree). However, the axiom of choice failed in their model. The present author later overcame this difficulty, but in the meantime Solovay had shown that if **ZF** has a model, then there is a model for **ZF** $+ V$ "is the constructible closure of a real which is the unique solution of a Π_2^1 predicate (hence is Δ_3^1)". In fact, by combining the techniques of Sacks and Solovay, we can insure that the model contains a real which is Δ_3^1 *and* of minimal degree. However, all these methods of constructing definable reals depended essentially on adjoining to the initial model, L, sets having different degrees of constructibility. Thus the question remained open, whether V can be the constructible closure of an analytic set of minimal degree. We now provide an affirmative answer. The set of forcing conditions used in the proof is constructed in a manner analogous to our earlier construction of a Souslin tree in L.* The conditions have cardinality ω_1 and satisfy the countable chain condition (CCC) in L; hence we may state the result in the following absolute form:

---

* Cf. Souslin's Hypothesis is incompatible with $V = L$, by Ronald Jensen; to appear in *J. S. L.*

THEOREM. *If* $\omega_1 > \omega_1^L$, *then there is a set* $a \subset \omega$ *of minimal degree of constructibility such that* a *is the unique solution of a* $\mathbf{\Pi}_2^1$ *predicate in* $L[a]$.

The present paper is devoted to a proof of this theorem.

§2. We assume basic knowledge of forcing. However, since forcing is usually presented as an operation on **ZF** models, we restate some of the definitions in a more general setting. A set of *forcing conditions* is a pair $\mathbb{P} = \langle |\mathbb{P}|, \leqq \rangle$ s.t. $|\mathbb{P}|$ is a non-empty set and $\leqq$ is a partial ordering of $|\mathbb{P}|$. '$P \leqq Q$' is read 'P extends Q'. We call $P, Q \in \mathbb{P}$ *compatible* if they have a common extension. A set $\Delta \subset \mathbb{P}$ is called *dense* in $\mathbb{P}$ iff every $P \in \mathbb{P}$ has an extension in Δ. $X \subset \mathbb{P}$ is called *pre-dense* in $\mathbb{P}$ iff every $P \in \mathbb{P}$ is compatible with an element of X. Now let M be a transitive collection of sets s.t. $\mathbb{P} \in M$. We call a set $X \subset M$ *M-definable* iff X is definable in the model $\langle M, \varepsilon \rangle$, using parameters from M. We call $G \subset \mathbb{P}$ *$\mathbb{P}$-generic over M* iff

(i) Any two elements of G are compatible

(ii) $P \geqq P' \wedge P' \in G. \rightarrow P \in G$

(iii) $G \cap \Delta \neq \emptyset$ for every M-definable dense Δ.

Clearly we can replace (iii) in this definition by:

(iii)′ $G \cap X \neq \emptyset$ for every M-definable pre-dense X.

The main lemmas on generic sets read:

(a) If M is countable, then there exists a G which is $\mathbb{P}$-generic over M.

(b) If M is a **ZF** model, then so is $M[G]$.

In proving (a), we need countability only to enumerate the M-definable dense subsets of $\mathbb{P}$. In fact, it suffices to have an enumeration X_i $(i < \omega)$ of pre-dense subsets s.t. whenever Δ is an M-definable dense subset, then $\mathrm{V}_i X_i \subset \Delta$. Hence, we get:

(c) If M is a model of **ZF** $+ CH$, if $\mathbb{P}$ satisfies CCC and has cardinality $\leqq \omega_1$ in M, and if $\omega_1 > \omega_1^M$, then there is a G which is $\mathbb{P}$-generic over M.

We note, for later use, one further refinement: If the hypotheses of (c) are satisfied and if $X \subset \mathbb{P}$ is such that whenever Δ is M-definable and dense in $\mathbb{P}$, then $\Delta \cap X$ is dense in X, then there is a $G \subset X$ which is $\mathbb{P}$-generic over M. Let us note, finally, that from sets of forcing conditions $\mathbb{P}_1, \cdots, \mathbb{P}_n$ we can form the cartesian product, giving it the componentwise partial ordering. The *product lemma* says that $G_1 \times G_2$ is $\mathbb{P}_1 \times \mathbb{P}_2$-generic over a **ZF** model M iff G_1 is $\mathbb{P}_1$-generic over M and G_2 is $\mathbb{P}_2$-generic over $M[G_1]$. In this case, we also have:

$$M[G_1] \cap M[G_2] = M.$$

Cohen's original model may be obtained by using as conditions the collection $2^{<\omega}$ of all finite 0, 1-sequences. The extension relation is the inverse of inclusion. Note that $2^{<\omega}$, ordered by inclusion, is a well-founded binary tree of length ω, Sacks modified Cohen's forcing by adopting as conditions the set of all *perfect subtrees* of $2^{<\omega}$. We define a subtree T of $2^{<\omega}$ to be perfect iff $T \neq \emptyset$ and for every $u \in T$ there are $v, v' \in T$ s.t. $v, v' \supset u$ and $v \mid v'$ in T. The extension relation coincides with inclusion. From now on, we take '$\leqq$' as denoting the inclusion relation on perfect trees. We say that a set $a \subset \omega$ *goes through* T iff the characteristic function of a is the union of a branch through T. Note that T is perfect iff the set $\{a \subset \omega \mid a \text{ goes through } T\}$ is a closed perfect set of reals. If T is perfect and $s \in T$, we set:

$$T^s = \{u \in T \mid u \subset s \wedge s \subset u\}.$$

T^s is then perfect. Now let $\mathbb{P}$ be a collection of perfect trees s.t.

$$T \in \mathbb{P} \wedge s \in T. \rightarrow T^s \in \mathbb{P}.$$

Let $\mathbb{P} \in M$, where M is a transitive collection which contains all hereditarily finite sets. If G is $\mathbb{P}$-generic over M, then there is exactly one $a \subset \omega$ which goes through each $T \in G$. Moreover, we can define G from a by:

$$G = G_a =_{Df} \{T \in \mathbb{P} \mid a \text{ goes through } T\}.$$

Hence $M[G] = M[a]$. Thus we can speak of generic reals rather than generic subsets of $\mathbb{P}$; in general, we define $a \subset \omega$ to be *$\mathbb{P}$-generic over M* iff G_a is $\mathbb{P}$-generic over M. Similarly, we define $\langle a_1, \ldots, a_n \rangle$ to be $\mathbb{P}^n$-generic iff $X_{i=1}^n G_{ai}$ is $\mathbb{P}^n$-generic. Sacks proved that if M is a model of $\mathbf{ZF} + V = L$, if $\mathbb{P}$ is the collection of all perfect trees in M, and if $a \subset \omega$ is $\mathbb{P}$-generic, then a has minimal degree in $M[a]$. His major tool is the *fusion lemma*, which says that if $\langle T_s \mid s \in 2^{<\omega} \rangle$ is a sequence of perfect trees s.t.

(i) $s \subset s' \rightarrow T_s \geqq T_{s'}$

(ii) If $f \in 2^\omega$, then $\bigcap_{n<\omega} T_{f \restriction n}$ contains only one branch

(iii) If $f, f' \in 2^\omega$, $f \neq f'$, then $\bigcap_n T_{f \restriction n} \neq \bigcap_n T_{f' \restriction n}$,

then $T^* = \bigcap_n \bigcup_{|s|=n} T_s$ is a perfect tree. (Here $|s|$ denotes the length of the sequence s.) Let us write $S \| T$ to mean: S, T are perfect trees with only a finite stem in common; i.e.

$$\mathrm{V}_s \in 2^{<\omega} \wedge u \in S \cap T \; u \subset s.$$

Clearly, the hypotheses of the fusion lemma are satisfied if there are arbitrarily large $n < \omega$ s.t.

$$|s| = |s'| = n,\ s \neq s' \rightarrow T_s \| T_{s'}.$$

§3. Our own conditions $\mathbb{P}$ will form a proper subset of the collection of all constructible perfect trees. We construct $\mathbb{P}$ in L, setting $\mathbb{P} = \bigcup_{\alpha<\omega_1} \mathbb{P}_\alpha$ where $\mathbb{P}_\alpha$ ($\alpha < \omega_1$) is a sequence of countable sets in L. We define $\mathbb{P}_\alpha$ by induction on α as follows:

$$\mathbb{P}_0 = \{(2^{<\omega})^u \mid u \in 2^{<\omega}\}$$

$$\mathbb{P}_\lambda = \bigcup_{\nu<\lambda} \mathbb{P}_\nu \text{ for limit } \lambda$$

We construct $\mathbb{P}_{\alpha+1}$ by forcing over L_γ where $\gamma = \gamma_\alpha$ is the least ordinal s.t. $\mathbb{P}_\alpha \in L_\gamma$ and

$$\mathfrak{P}(\omega) \cap L_{\gamma+1} \not\subset L_\gamma .$$

As conditions, we use the collection $\mathbb{C}$ of all finite maps P from a subset of $\omega \times 2^{<\omega}$ to $\mathbb{P}_\alpha$ s.t. whenever $u, v \in 2^{<\omega}$, $u \subset v$ and P_{nv} is defined, then P_{nu} is defined and $P_{nu} \geqq P_{nv}$. The extension relation on $\mathbb{C}$ is defined by:

$$P \leqq Q \leftrightarrow_{Df} \quad Q = P \restriction \mathrm{dom}(Q),$$

Since L_γ is countable, there exists a $\mathbb{C}$-generic subset $\mathfrak{S}$ of $\mathbb{C}$. Let $\mathfrak{S}^{(\alpha)}$ be the least such (in the canonical well ordering of L) and set:

$$S = S^{(\alpha)} =_{Df} \bigcup \mathfrak{S}^{(\alpha)} .$$

Then S is a map from $\omega \times 2^{<\omega}$ to $\mathbb{P}_\alpha$. By $\mathbb{C}$-genericity, for each $h < \omega$ there are arbitrarily large $n < \omega$ s.t.

$$|u| = |v| = n \wedge u \neq v . \rightarrow S_{hu} \,||\, S_{hv} .$$

Hence, by the fusion lemma, we may define:

$$S_h = S_h^{(\alpha)} =_{Df} \bigcap_n \bigcup_{|u|=n} S_{hu} \qquad (h < \omega),$$

obtaining a new sequence of perfect trees. We set:

$$\mathbb{P}_{\alpha+1} = \mathbb{P}_\alpha \cup \{S_n^v \mid n < \omega,\ v \in S_n\} .$$

The set of conditions $\mathbb{P} = \bigcup_\alpha \mathbb{P}_\alpha$ was defined in analogy with the author's earlier construction of a Souslin tree in L. Hence, we may expect it to have roughly analogous properties. In particular, we may hope that it satisfies CCC in L. We shall show that this is so and, in fact, that the cartesian product $\mathbb{P}^n$ satisfies CCC in L. We begin with a technical lemma:

LEMMA 1. *If $\Delta \subset \mathbb{P}_\alpha^n$ is L_{γ_α}-definable and dense and closed under extension in $\mathbb{P}_\alpha^n$, then for $k, m_1, \ldots, m_n < \omega$ there exists $h \geqq k$ s.t. if $s_1, \ldots, s_n \in 2^{<\omega}$, $|s_i| = h$ and $s_i \neq s_j$ for $i \neq j$, then*

$$\langle S_{m_1 s_1}^{(\alpha)}, \ldots, S_{m_n s_n}^{(\alpha)} \rangle \in \Delta .$$

Lemma 1 follows immediately from the $\mathbb{C}$-genericity of $S^{(\alpha)}$,

COROLLARY 2. *If* $b_i \subset \omega$ *goes through* $S^{(\alpha)}_{m_i}$ $(i = 1, \dots, n)$ *and* $b_1, \dots, b_n$ *are distinct, then* $\langle b_1, \dots, b_n \rangle$ *is* $\mathbb{P}^n_\alpha$*-generic over* L_{γ_α}.

By Corollary 2, no element of $\mathbb{P}_{\alpha+1} \setminus \mathbb{P}_\alpha$ is an element of L_{γ_α}. But for $\beta > \alpha$ we have: $\mathbb{P}_{\alpha+1} \subset \mathbb{P}_\beta \in L_{\gamma_\beta}$; hence:

$$\alpha < \beta \rightarrow \gamma_\alpha < \gamma_\beta .$$

LEMMA 3. *If* $X \subset \mathbb{P}^n_\alpha$ *is* L_{γ_α}*-definable and pre-dense in* $\mathbb{P}^n_\alpha$*, then* X *is pre-dense in* $\mathbb{P}^n_{\alpha+1}$.

Proof. We note first that $\mathbb{P}_{\alpha+1} \setminus \mathbb{P}_\alpha$ is dense in $\mathbb{P}_{\alpha+1}$ since, by $\mathbb{C}$-genericity, $\{S_{n\phi} \mid n < \omega\}$ is dense in $\mathbb{P}_\alpha$ and each $P \in \mathbb{P}_{\alpha+1} \setminus \mathbb{P}_\alpha$ extends some $S_{n\phi}$. Thus, we need only show that each $P \in (\mathbb{P}_{\alpha+1} \setminus \mathbb{P}_\alpha)^n$ is compatible with an element of X. Let $P = \langle S^{u_1}_{m_1}, \dots, S^{u_n}_{m_n} \rangle$. We must find a Q which is a common extension of P and some element of X. Set:

$$\bar{X} = \{R \in \mathbb{P}^n_\alpha \mid \bigvee Q \in X \ R \leqq Q\}.$$

$\bar{X}$ is L_{γ_α}-definable and dense in $\mathbb{P}_\alpha$. Let $|u_i| < k \ (i = 1, \dots, n)$. By Lemma 1, there is $h \geqq k$ s.t. if $v_1 \dots, v_n \in 2^{<\omega}$, $|v_i| = h$, $v_i \neq v_j$ $(i \neq j)$, then $\langle S_{m_1 v_1}, \dots, S_{m_n v_n} \rangle \in \bar{X}$. Pick $v_1, \dots, v_n$ satisfying these conditions s.t. $v_i \supset u_i$ $(i = 1, \dots, n)$. Then $Q = \langle S^{v_1}_{m_1}, \dots, S^{v_n}_{m_n} \rangle$ fulfills the requirement. Q.E.D.

COROLLARY 4. *If* $X \subset \mathbb{P}^n_\alpha$ *is* L_{γ_α}*-definable and pre-dense in* $\mathbb{P}^n_\alpha$*, then* X *is pre-dense in* $\mathbb{P}^n$.

Proof. It suffices to show that X is pre-dense in $\mathbb{P}_\beta$ for $\alpha \leqq \beta < \omega_1$. We prove this by induction on β, using Lemma 3 and the monotonicity of γ_β. Q.E.D.

COROLLARY 5. $\{S^{(\alpha)}_n \mid n < \omega\}$ *is pre-dense in* $\mathbb{P}$.

Proof. Since each element of $\mathbb{P}_{\alpha+1} \setminus \mathbb{P}_\alpha$ extends some $S^{(\alpha)}_n$, it suffices to show that $\mathbb{P}_{\alpha+1} \setminus \mathbb{P}_\alpha$ is pre-dense in $\mathbb{P}$. $\mathbb{P}_{\alpha+1} \setminus \mathbb{P}_\alpha$ is $L_{\gamma_{\alpha+1}}$-definable. Moreover, $\mathbb{P}_{\alpha+1} \setminus \mathbb{P}_\alpha$ is dense in $\mathbb{P}_{\alpha+1}$, as noted in the proof of Lemma 3. The conclusion follows by Corollary 4. Q.E.D.

We are now ready to prove:

LEMMA 6. $\mathbb{P}^n$ *satisfies CCC in L.*

Proof. Assume $V = L$. Let $X \subset \mathbb{P}_n$ be a maximal set of pairwise incompatible conditions. We must show that X is countable. $\mathbb{P}, X$ are subsets of L_{ω_1}, hence $\mathbb{P}, X \in L_{\omega_2}$. Let M be a countable elementary submodel of $\langle L_{\omega_2}, \varepsilon \rangle$ s.t. $\mathbb{P}, X \in M$. By the condensation lemma, there exists a map:

$$\pi\colon M \leftrightarrow \langle L_\beta, \varepsilon\rangle$$

for some countable β. Let α be the least ordinal not an element of M. Then:

$$\alpha = \pi(\omega_1)$$

$$y \in M,\ \varepsilon L_{\omega_1} \to \pi(y) = y$$

$$Y \in M,\ \subset L_{\omega_1} \to \pi(Y) = Y \cap L_\alpha .$$

In particular, since $\mathbb{P} = \bigcup_{\nu<\omega_1} \mathbb{P}_\nu$ and $\mathbb{P}_\nu \in L_{\omega_1}$ for $\nu < \omega_1$, we have:

$$\pi(\mathbb{P}) = \bigcup_{\nu<\alpha} \mathbb{P}_\nu = \mathbb{P}_\alpha .$$

By definition, γ_α is the least γ s.t. $\mathbb{P}_\alpha \in L_\gamma$ and $\mathfrak{P}(\omega) \cap L_{\gamma+1} \not\subset L_\gamma$. Moreover, $\gamma_\alpha \geqq \alpha$. Since $\beta > \alpha$ and $\mathfrak{P}(\omega) \cap L_\beta \subset L_\alpha$, we may conclude that $\beta \leqq \gamma_\alpha$. Hence

$$\pi(X) = X \cap \mathbb{P}^n_\alpha \in L_\beta \subset L_{\gamma_\alpha} .$$

Since X is pre-dense in $\mathbb{P}$, $\pi(X) = X \cap \mathbb{P}^n_\alpha$ is pre-dense in $\mathbb{P}^n_\alpha$. By Corollary 4, then, $X \cap \mathbb{P}^n_\alpha$ is pre-dense in $\mathbb{P}$. Since X is a set of pairwise incompatible conditions, this means that $X = X \cap \mathbb{P}^n_\alpha$. Hence X is countable.

Q.E.D.

Putting our lemmas together, we obtain a very simple characterisation of $\mathbb{P}^n$-generic n-tuples:

LEMMA 7. *Let $b_1, \ldots, b_n \subset \omega$. Then $\langle b_1, \ldots, b_n\rangle$ is $\mathbb{P}^n$-generic over L iff $b_1, \ldots, b_n$ are distinct and*

$$\bigwedge \alpha \bigvee h\ (b_i \text{ goes through } S_h^{(\alpha)})$$

for $i = 1, \ldots, n$.

Proof. The necessity of the condition is obvious by Corollary 5. We now prove sufficiency. Let $\langle b_1, \ldots, b_n\rangle$ satisfy the condition and let $\Delta \subset \mathbb{P}_n$ be dense. We must show that:

$$\left(\mathop{\times}_{i=1}^{n} G_{b_i}\right) \cap \Delta \neq \emptyset .$$

Let $X \subset \Delta$ be a maximal set of pairwise incompatible conditions. Since X is countable, there is an α s.t. $X \in \mathbb{P}_\alpha$ and X is L_{γ_α}-definable. Since X is pre-dense in $\mathbb{P}_\alpha$, the conclusion follows by Corollary 2. Q.E.D.

Immediate corollaries of Lemma 7 are:

COROLLARY 8. *If a, b are $\mathbb{P}$-generic over L and $a \neq b$, then $\langle a, b\rangle$ is $\mathbb{P}^2$-generic over L.*

COROLLARY 9. *Let M be an inner* **ZF** *model s.t. $L \subset M$. Then the set $A =_{Df} \{a \subset \omega \mid a$ is $\mathbb{P}$-generic over $L\}$ is Π^1_2 in M.*

Proof. By a theorem of Takeuti, it suffices to show that A is Π_1 in the collection H_{ω_1} of hereditarily countable sets. Let $\tau = \omega_1^L$. It is easily seen that the function

$$\langle S_n^{(\alpha)} \mid \alpha < \tau,\ n < \omega \rangle$$

is Δ_1 (i.e. recursive) in L_τ. Since τ, L_τ are Σ_1 in H_{ω_1}, this function is Δ_1 in H_{ω_1}. But

$$a \in A \leftrightarrow \bigwedge \alpha < \tau \bigvee n \ (a \text{ goes through } S_n^{(\alpha)})$$

Q.E.D.

The next two lemmas, which follow directly from Corollary 8, establish the theorem.

LEMMA 10. *If a is $\mathbb{P}$-generic over L, then $A = \{a\}$ in $L[a]$.*

Proof. Let $b \neq a$ be $\mathbb{P}$-generic over L. Then $b \notin L[a]$, since by Corollary 8 $\langle b, a \rangle$ is $\mathbb{P}^2$-generic over L; i.e. b is $\mathbb{P}$-generic over $L[a]$. Q.E.D.

LEMMA 11. *If a is $\mathbb{P}$-generic over L, then a is of minimal degree.*

Proof. It suffices to prove the lemma under the assumption $\omega_1 > \omega_1^L$. Let x be a set of ordinals s.t. $x \in L[a]$ but $a \notin L[x]$. We wish to prove: $x \in L$. Let X be the set of conditions compatible with the interpretation x of $\dot{x}$; i.e. X is the set of all $P \in \mathbb{P}$ which do not force the negation of any true bounded quantifier statement about $L[x]$. Obviously $L[x] = L[X]$. A simple forcing argument shows that if $\Delta \subset \mathbb{P}$, $\Delta \in L$ and Δ is dense in $\mathbb{P}$, then $\Delta \cap X$ is dense in X. By this it follows easily, using $\omega_1 > \omega_1^L$, that if $P \in X$, then $P \in G_b \subset X$ for some b which is $\mathbb{P}$-generic over L. By definition, $G_a \subset X$. But $G_a \neq X$, since otherwise $a \in L[X] = L[x]$. Hence there must be a $\mathbb{P}$-generic b s.t. $b \neq a$ and $G_b \subset X$. Then $\dot{x}$ denotes x in $L[b]$. Since $\langle a, b \rangle$ is $\mathbb{P}^2$-generic, we conclude:

$$x \in L[a] \cap L[b] \subset L$$

Q.E.D.

DEFINABILITY IN AXIOMATIC SET THEORY II*

BY

AZRIEL LEVY

Hebrew University of Jerusalem

§ 1. Introduction and statement of the results. Let **ZF** be the Zermelo-Fraenkel set theory, and let **ZFC** be **ZF** with the axiom of choice added. Let **GCH** denote the generalized continuum hypothesis. We shall refer to subsets of ω as real numbers, but we shall do it only in those cases where what we say is indeed equally true for the genuine real numbers.

THEOREM 1 ([5, Th. 1]). *If* **ZF** *is consistent, so is* **ZFC** + **GCH** *together with the following additional axioms*:

(1) *There exists a nonconstructible real number.*

(2) *Every hereditarily-ordinal-definable set is constructible.*

(3) *There is a real number a such that* $V = L[a]$, i.e., *every set is constructible from a.*

(4) *Every constructible cardinal is a true cardinal, or, in other words, if* $\alpha < \beta$ *are ordinals such that there is a mapping of* α *on* β *then there is also a* constructible *mapping of* α *on* β.

Theorem 1 of [5] mentions only (1) and (2). (3) follows directly from the construction of the model N of **ZF** + **GCH** + (1) + (2). (4) is shown to hold in that model by the methods of Cohen [2, p. 132].

(3) implies the **GCH** in **ZF** (see Shoenfield [9, p. 539]). As shown in [5], (1) and (2) imply in **ZFC**:

(5) There is no ordinal-definable well-ordering of all real numbers; and

(6) There is a Π_1^1-predicate $P(f)$ of number theory such that there is a function $f \in \omega^\omega$ which satisfies it, but no such function is ordinal-definable.

(5) asserts that there is no ordinal-definable well-ordering of *all* real

* The present paper is a continuation of [5]. In [5] proofs were given of results announced in [6]. The present paper contains proofs of the results of Feferman and the author in [3] and [6], and some additional results. We make use of new formulations and results which were published in the meanwhile, in particular those of Solovay [11]. The author is deeply indebted to John W. Addison, Paul J. Cohen, Solomon Feferman, Dana Scott, Robert M. Solovay and Robert L. Vaught for many stimulating and helpful discussions of the problems handled in the present paper. This work was supported by the United States Office for Naval Research, Information Systems Branch, Contract No. N00014 69 C 0192.

numbers. However, there is a definable, and even a Δ^1_2, well-ordering of all constructible real numbers. By (4) and the **GCH** the set of all constructible real numbers is of cardinality $\aleph_1 = 2^{\aleph_0}$. Therefore, if we want to strengthen (5) to

(7) Every ordinal-definable well-ordering of real numbers is countable, then we must give up (4).

THEOREM 2. *If* **ZF** *is consistent, so is* **ZFC** + **GCH** *together with* (2), (3) *and* (7).

In **ZFC** (7) implies (1) which, together with (2), implies (6).

(6) says that even the set of all ordinal-definable real numbers is not a basis for Π^1_2. A closely related problem is the question whether the Novikoff-Kondo-Addison uniformization theorem ([10, p. 188]) can be extended to Π^1_2-relations. This uniformization theorem says that if $P(f, g)$ is a Π^1_1-relation then it has a Π^1_1-subrelation $Q(f, g)$ with the same domain such that Q is a function, i.e., for every f there is at most one g such that $Q(f, g)$. In this case we say that the relation P is uniformized by Q. It is a simple consequence of (6) that there is a Π^1_2-relation $R(f, g)$ which cannot be uniformized by any ordinal-definable relation, as we shall now see. Let $P(f)$ be as in (6) and let $R(f, g)$ be the relation given by $f = 0 \wedge P(g)$. Suppose that $Q(f, g)$ is an ordinal-definable relation which uniformizes $R(f, g)$; then the range of R consists of a single function g such that $P(g)$, and this function g is obviously ordinal-definable (by the way in which we described it), contradicting (6).

Even though (6) implies that there is a Π^1_2-relation which cannot be uniformized by an ordinal-definable relation, (3) implies that every ordinal-definable relation can be uniformized by a relation ordinal-definable in some real number. If a is a real number such that $V = L[a]$, then we have a natural well-ordering of all sets which is definable in a. An ordinal-definable relation $P(f, g)$ on ω^ω is uniformized by the relation $Q(f, g)$ given by "$P(f, g)$, and for no $g' \in \omega^\omega$ which precedes g in the natural ordering of $L[a]$ does $P(f, g')$ hold", which is ordinal-definable in a. In fact, if P is a Π^1_2-relation it follows from (3) that it can be uniformized by a relation Δ^1_3 in a. Therefore we must give up (3) in order to obtain

(8) There is a Π^1_2-relation which cannot be uniformized by any real-ordinal-definable relation (i.e., by a relation ordinal-definable from some real number).

THEOREM 3. *If* **ZF** *is consistent, so is* **ZFC** + **GCH** *together with* (8).

(8) implies also, by the same argument we used above to prove that (3) is incompatible with (8), that

(9) There is no real-ordinal-definable well-ordering of the set of all real numbers.

A common strengthening of (7) and (9) is

(10) Every real-ordinal-definable well-ordering of real numbers is countable.

In **ZFC** + (10) we can prove

(11) ω_1 is an inaccessible number in the constructible universe.

To prove this, let ω_α^L denote the αth infinite constructible cardinal (where we start counting with 0). Let $\omega_1 = \omega_\xi^L$. Since ω_1 is regular it is also regular in the constructible universe, therefore (11) will follow once we prove that ξ is a limit ordinal. Suppose $\xi = \eta + 1$ for some ordinal η, then $\omega_\eta^L < \omega_\xi^L = \omega_1$, hence ω_η^L is countable. Therefore there is a real r which codes a well-ordering of ω of the order-type ω_η^L (i.e., if R is a binary relation well-ordering ω in the order-type ω_η^L then $n \in r \leftrightarrow \exists k \exists l\, (kRl \wedge n = 2^k \cdot 3^l)$). In the universe $L[r]$ of all sets constructible from r, ω_η^L is denumerable since r codes a well-ordering of ω of this order-type . ω_1, being a true cardinal, is also a cardinal of $L[r]$. Since even L has no cardinal between ω_η^L and ω_1 and since ω_η^L is countable in $L[r]$ we have $\omega_1 = \omega_1^{L[r]}$. Since Cantor's theorem holds in $L[r]$ there are at least $\omega_1^{L[r]} = \omega_1$ real numbers in $L[r]$. These real numbers can be well-ordered by a well-ordering which is definable from r, which contradicts (10).

Therefore, if we prove the consistency of **ZFC** + (10) we will also have established, once we go over to the model L of Gödel, and since (10) implies (11), that the set theory **ZFCI** is consistent, where **ZFCI** is the set theory obtained from **ZFC** by adding an axiom which asserts the existence of inaccessible cardinals. It is well known (see, e.g. [10, p. 306]), that it follows from Gödel's theorem on consistency proofs that if **ZF** is consistent then we cannot prove the statement "if **ZF** is consistent then so is also **ZFCI**". Thus, in order to prove the consistency of **ZFC** + (10) it is not enough to assume the consistency of **ZF**. We shall, therefore, assume the consistency of **ZFCI**.

THEOREM 4. *If* **ZFCI** (*i.e.*, **ZFC** *together with the existence of an inaccessible number*) *is consistent, then so is* **ZFC** + (10).

Another question with which we shall deal is the following. Suppose we do not assume the **GCH**, and suppose $\aleph_0 < \aleph_\alpha < 2^{\aleph_0}$; is there a *definable* set of real numbers of cardinality $\aleph_\alpha$? We can prove:

THEOREM 5. *If* Λ *is a suitably defined ordinal* (see, e.g., *the absolutely definable ordinals of Hajnal* [4, p. 324]), *if* **ZF** *is consistent and*

"$\Lambda > 0 \wedge \Lambda$ *is not confinal with* ω" *is consistent with* **ZF**, *then* **ZFC** *is consistent with* $2^{\aleph_0} = \aleph_\Lambda$, *together with* (4) *and*

(12) *Every real-ordinal-definable set of real numbers has cardinality* $\leqq \aleph_1$ *or* $2^{\aleph_0}$.

(4) implies that the set of all constructible reals, which is obviously a definable set, is of cardinality $\aleph_1$. By dropping (4) we can replace (12) by the stronger statement (13).

(13) Every ordinal-definable set of real values is of cardinality $\leqq \aleph_0$ or $2^{\aleph_0}$.

THEOREM 6. *If* Λ *is a suitably defined ordinal, if* **ZF** *is consistent and* "$\Lambda > 0 \wedge \Lambda$ *is not confinal with* ω" *is consistent with* **ZF**, *then* **ZFC** *is consistent with* $2^{\aleph_0} = \aleph_\Lambda$ *together with* (13).

It is now natural to ask whether we can improve Theorem 6 by replacing in (13) "ordinal-definable" by "real-ordinal-definable", thereby obtaining

(14) Every real-ordinal definable set of real numbers is of cardinality $\leqq \aleph_0$ or $2^{\aleph_0}$.

It turns out that in **ZFC** together with $2^{\aleph_0} > \aleph_1$ and (14) one can prove (11). To do this, all we have to do, as we did when we showed that (10) implies (11) in **ZFC**, is to show that $\omega_1 \neq \omega^L_{\eta+1}$, for every ordinal η. If $\omega_1 = \omega^L_{\eta+1}$ then ω^L_η is countable and let r be a real which codes an ordering of ω of order-type ω^L_η. It is easily seen that the set T of all reals constructible from r is of the cardinality $\aleph_1$. T is a real ordinal-definable set, contradicting (14). Since we have obtained (11), we must, as we did in Theorem 4, strengthen our assumption in Theorem 6 if we want to replace (13) with (14). In fact, Solovay has obtained in [12] the even stronger result:

THEOREM 7 (SOLOVAY [11]). *If* **ZFCI** (*i.e.*, **ZFC** *with the existence of an inaccessible number*) *is consistent, and it is also consistent with* "$\Lambda > 0$ *and* Λ *is not confinal with* ω", *where* Λ *is a suitably defined ordinal, then so is* **ZFC** *together with* $2^{\aleph_0} = \aleph_\Lambda$ *and*

(15) *Every real-ordinal-definable set is of cardinality* $\leqq \aleph_0$ *or includes a perfect set* (*which is of cardinality* $2^{\aleph_0}$).

By methods similar to those used in the proofs of Theorems 1–4 we have proved, jointly with S. Feferman, the following Theorem 8. Its proof is given in [2, Ch. IV, §10].

THEOREM 8. *If* **ZF** *is consistent then it is consistent with*

(16) *The continuum is a countable union of countable sets*, and

(17) $\omega_1 = \omega_\omega^L$ (*where* ω_α^L *is the* α*-th infinite constructible cardinal*)

As shown by Specker [12, III, §3], (16) has many interesting consequences, in particular:

(18) ω_1 is a countable union of countable sets, or, equivalently

(19) there is an increasing sequence α_n of ordinals such that $\lim_{n<\omega}\alpha_n = \omega_1$.

(18) and (19) follow also, obviously, from (17). Another interesting consequence of (17), which we shall prove below, is

(20) There is a Π_2^1-relation $R(n,f)$ such that $(\forall n \in \omega)(\exists f \in \omega^\omega)R(n,f)$, but there is no sequence $\langle f_0, f_1, \ldots\rangle$ such that $(\forall n \in \omega) R(n,f_n)$.

(20) is the strongest possible failure of the axiom of choice from the point of view of the theory of definability. For a Π_1^1-relation, and even for a Σ_2^1-relation $R(n,f)$, we can prove, by means of the Novikoff-Kondo-Addison uniformization theorem ([10, p. 188]), that $(\forall n \in \omega)(\exists f \in \omega^\omega) R(n,f)$ implies the existence of a sequence $f_0, f_1, \ldots$ such that $(\forall n \in \omega) R(n,f_n)$. The uniformization theorem is proved without using the axiom of choice, and thus for these simple cases, the axiom can be proved rather than postulated.

We shall conclude this section by showing that (17) implies (20). For $f \in \omega^\omega$ we shall say "*f is an ordering of natural numbers*" if the relation $\{\langle k,l\rangle \mid f(2^k \cdot 3^l) = 0\}$ is an ordering of a subset of ω, and we shall also call that ordering f. For $R(n,f)$ we shall take an analytical formula which is equivalent to the statement "*f is a well-ordering of order-type* $\geqq \omega_n^L$". Let us show first that the axiom of choice fails for such a formula, and then we shall prove that it is a Π_2^1-formula. Given any $n < \omega$, we have $\omega_n^L < \omega_\omega^L = \omega_1$, hence ω_n^L is countable, and therefore there is a well-ordering of ω of order-type $\geqq \omega_n^L$. Suppose there is a sequence $\langle f_0, f_1, \ldots\rangle$ such that for $n < \omega$, f_n is a well-ordering of natural numbers of order-type $\geqq \omega_n^L$. Then the relation $\{\langle 2^m \cdot 3^k, 2^n \cdot 3^l\rangle \mid m < n \text{ or } m = n \text{ and } f_n(2^k \cdot 3^l) = 0\}$ is obviously a well-ordering of natural numbers; let us denote its order-type with α. α is not less than ω_n^L (since the order-type of the subset of the field of that relation which consists of the numbers $2^n \cdot 3^k$, for $k < \omega$, is $\geqq \omega_n^L$). Since $\alpha \geqq \omega_n^L$ for every n we have $\alpha \geqq \sup_{n<\omega}\omega_n^L = \omega_\omega^L = \omega_1$. This is a contradiction since no countable set can be of order-type $\geqq \omega_1$.

It is well known, that there is a single sentence ψ of **ZF**, such that for every transitive set T, ψ is true in $\langle T, \in\rangle$, if and only if $T = L_\gamma$, for some infinite limit number γ, where L_γ is the set of all constructible sets, obtained by less than γ steps, in the process of constructing the subsets of the hitherto defined universe, which are definable in terms of the already defined sets. (L_γ is the set M_γ of Cohen [2, Ch. III].)

To compute the place of $R(n,f)$, in the analytical hierarchy, let us pass from "*f is an ordering of natural numbers of order-type* $\geqq \omega_n^L$", to the equivalent statement

(21) *f is a well-ordering of natural numbers, whose order-type we denote with* α, *and every countable transitive model* M *of* ψ, *which contains the ordinal* α, *has* $n+1$ *infinite ordinals* $\leqq \alpha$, *which are cardinals in the sense of* M.

Let us prove first the equivalence of the two statements. Assume that the order-type α of f is $\geqq \omega_n^L$. Let M be a countable transitive model of ψ. By the characteristic property of ψ mentioned above, $M = L_\gamma$ for some infinite number γ, and hence all the numbers of M are constructible sets. Therefore every constructible cardinal ω_β^L which belongs to M, is also a cardinal in the sense of M, because if there is no constructible function mapping an ordinal $\delta < \omega_\beta^L$ on ω_β^L, there is no such function in M either. Thus the $n+1$ infinite constructible cardinals $\omega_0^L, \omega_1^L, \dots, \omega_n^L \leqq \alpha$, are also cardinals in the sense of M, and (21) holds. To prove the other direction, let us assume that (21) holds. Since α is countable, $\alpha < \omega_1$ and hence $\alpha < \omega_m^L$ for some $0 < m < \omega$. Let $\beta < \omega_m^L$. If β is not a constructible cardinal, then there is a constructible one-one function g mapping β on some ordinal $\gamma < \beta$. Since $g \subseteq \beta \times \gamma$, and $\gamma < \beta < \omega_m^L$, one can show that $g \in \omega_m^L$ by the same arguments used to prove that, for an infinite β, if $x \subseteq L_\beta$ and $y \leqq x$, then $y \in L_\delta$ for some ordinal δ such that $|\delta| \leqq |\beta|$ (see, e.g., Cohen [2, Ch. II, §4, Theorem 1]). Thus if $\beta < \omega_m^L$ is not a constructible cardinal, it is also not a cardinal in the sense of $L_{\omega_m^L}$. Since ω_m^L is an infinite limit number, $L_{\omega_m^L}$ is a model of ψ. Therefore, by (21), $L_{\omega_m^L}$ has at least $n+1$ infinite ordinals $\leqq \alpha$ which are cardinals in the sense of $L_{\omega_m^L}$, and are hence constructible cardinals. This means that $\alpha \geqq \omega_m^L$, which is what we had to prove.

By the Mostowski-Shepherdson isomorphism theorem (e.g., Cohen [2, p. 73]) we can replace in (21), the notion of a countable transitive model of ψ, by that of a well-founded model of $\sigma \wedge \psi$, where σ is the axiom of extensionality. Thus (21) is equivalent to

(22) $\forall g, h \in \omega^\omega$ [f is a well-ordering of natural numbers $\wedge$ g codes a well-founded binary relation $\in'$ on a set $M \subseteq \omega \wedge \sigma \wedge \psi$ holds in the model $\langle M, \in' \rangle \wedge (\forall k \in M)$ (k is an ordinal in $\langle M, \in' \rangle \wedge h$ is an isomorphism of the field of the well-ordering f on the ordinals of $\langle M, \in' \rangle$ which stand in the relation $\in'$ to $k \rightarrow$ there are $n+1$ distinct ordinals $l_0, \dots, l_n$ of $\langle M, \in' \rangle$ such that for each $i \leqq n$, $l_i \in' k$ and l_i is an infinite cardinal of $\langle M, \in' \rangle$)].

In order to show that (22) is a Π_2^1-statement, it is enough to show that the part inside the square brackets, is a Σ_1^1-statement, and this is established once we show that inside the square brackets in (22), all the statements to the left of $\rightarrow$ are Π_1^1, while all the statements to the right of $\rightarrow$ are Σ_1^1. The statements "f is a well-ordering of natural numbers", and "g codes a well-founded binary relation", are well known, and easily shown to be Π_1^1. All other statements inside the square brackets of (22) are first-order statements with respect to $\langle M, \in' \rangle$ and are, hence, arithmetical statements, and, as such, are in $\Delta_1^1 = \Sigma_1^1 \cap \Pi_1^1$.

§ 2. Proof of Theorem 2. A very elegant and general exposition of forcing is given by Solovay in [11]. We shall use his way of looking at things, and some of his results, and we shall assume that the reader has read the first two sections of [11].

The general situation is as follows. There is a countable transitive model M of **ZF**. We shall denote the least ordinal not in M by α_0. In M, there is a partially ordered set, P whose members will be called *conditions*. This partial ordering of P will be denoted with $\leqq$ or $\leqq_P$. We assume that P has a minimum and denote it with 0. We shall often require that the ordering $\leqq_P$ satisfy

(23) For all $p, p' \in P$ there is in M an automorphism π of P, such that $\pi(p)$ is compatible with p' (i.e., there is a $q \in P$ such that $\pi(p), p' \leqq q$).

Since M is countable, there is an M-generic filter G over P. We extend M to a transitive model $M[G] = N$ of **ZFC**. We define a language L' which we intend to interpret in N. L' will be the language of **ZF**, with the following additional symbols.

(a) A unary predicate symbol S, such that $S(x)$ will be interpreted as $x \in M$.

(b) An individual constant G which will denote the M-generic filter G over P.

(c) An individual constant $\mathbf{x}$ for every member x of M.

In fact, $\langle N, \in \rangle$ satisfies all the instances of the axiom schema of replacement formulated in L', including those which contain the predicate symbol S. We shall denote weak forcing with $\Vdash$ (in [5] we denoted with $\Vdash$ strong forcing while weak forcing was denoted with $\Vdash^*$). We extend L' to another language L'' which has also a term for each set in N. Also, the language L'' is defined inside M. For the details of the structure of L'', see Cohen [2] or [5].

LEMMA 1. *Let Φ be a sentence of L' which does not contain G, and assume that* (23) *holds. Φ is true in N, if and only if,* $0 \Vdash \Phi$.

Proof. This is proved exactly like the proof of [11, Lemma I 3.5].

In the following, if $x_1, \ldots, x_n \in N$, we shall write $N \models \Phi[x_1, \ldots, x_n]$ if $x_1, \ldots, x_n$ satisfy the formula Φ in the structure $\langle N, \in \rangle$.

LEMMA 2. *Let $\Phi(x, \mathbf{y}_1, \ldots, \mathbf{y}_n)$ be a formula of L' which does not mention*

G with the only free variable x, and with no constants for members of M, except $\mathbf{y}_1, \ldots, \mathbf{y}_n$. *Let B be the class of N, such that for all x,* $x \in B \leftrightarrow x \in N \wedge N \vDash \Phi[x, y_1, \ldots, y_n]$. *If (23) holds then, there is a formula* $\Psi(x, y_1, \ldots, y_n, \leqq_P)$ *of set theory such that for the given* $y_1, \ldots, y_n, \leqq_P$, *and for all x*

$$x \in B \cap M \leftrightarrow x \in M \wedge M \vDash \Psi[x, y_1, \ldots, y_n, \leqq_P].$$

If, in addition, $B \in N$, *then also* $B \cap M \in M$.

Proof ([5, Lemma 62]).

$$B \cap M = \{x \mid x \in M \wedge N \vDash \Phi(x, \mathbf{y}_1, \ldots, \mathbf{y}_n)\}$$

$$= \{x \mid x \in M \wedge 0 \Vdash \Phi(\mathbf{x}, \mathbf{y}_1, \ldots, \mathbf{y}_n)\} \text{ (by Lemma 1)}$$

Since forcing is expressible in M there is a formula $\Psi(x, y_1, \ldots, y_n, z)$ of **ZF** such that for all $x \in M$

$$0 \Vdash \Phi(\mathbf{x}, \mathbf{y}_1, \ldots, \mathbf{y}_n) \leftrightarrow \Psi(x, y_1, \ldots, y_n, \leqq_P)$$

which proves the first claim of the theorem. If $B \in N$ then $B \subseteq R(\beta)$ for some $\beta < \alpha_0$, hence $B \cap M \subseteq R(\beta) \cap M = R(\beta)^{(M)}$, where $R(\beta)^{(M)}$ denotes the set $R(\beta)$ of M. Since the axiom of subsets holds in M, M contains the subset of $R(\beta)^{(M)}$, defined in M by the formula $\Psi(x, y_1, \ldots, y_n, \leqq_P)$, which is $B \cap M$.

LEMMA 3. *Assuming (23), then for every* $x \in N$ *and* $y \in M$, *if x is hereditarily-ordinal-definable from y in the sense of N then* $x \in M$.

Proof ([5, Lemma 63]). As shown by Myhill and Scott in [8], there is a formula $\Gamma(y, \alpha, x)$ of set theory, such that the formula $\exists \alpha \Gamma(y, \alpha, x)$ defines the property of being ordinal-definable from y, and such that

$$\mathbf{ZF} \vdash \Gamma(y, \alpha, x) \wedge \Gamma(y, \alpha, z) \rightarrow x = z.$$

Assume that there is a member s of $N - M$ which is hereditarily-ordinal-definable from y in the sense of N. By the axiom of foundation, and since the axiom of replacement holds in N for all formulas of L', there is such a set $s \subseteq M$. By what we said above there is an ordinal α in N, such that s is the only member of N for which $N \vDash \Gamma[y, \alpha, s]$. Since $\alpha \in N$, $\alpha < \alpha_0$. s can therefore be defined in N as a class of N by the formula

$$x \in s \leftrightarrow \exists z[x \in z \wedge \Gamma(\mathbf{y}, \boldsymbol{\alpha}, z)].$$

Since $y, \alpha \in M$ we get, by Lemma 2, $s = s \cap M \in M$, contradicting $s \in N - M$.

We proceed now to prove the theorem. Instead of dealing directly with consistency we shall start with a countable transitive model M of $\mathbf{ZF} + \mathbf{V} = \mathbf{L}$ and we shall construct a model of $\mathbf{ZFC} + \mathbf{GCH} + (3) + (7)$. This yields a proof of Theorem 2 by the methods given in Cohen [2, pp. 147–148], or in [5, pp. 132–133].

Now, M is a denumerable transitive model of $\mathbf{ZF} + \mathbf{V} = \mathbf{L}$. We take for P the usual set of conditions used to map ω on $\omega_1^{(M)}$ (Cohen [2, p. 143], or [11, I, 1.12]), namely P is the set of all functions on finite subsets of ω into $\omega_1^{(M)}$ ($\omega_1^{(M)}$ is the ω_1 of the model M). $\leqq_P$ is taken to be the inclusion relation. Let G be an M-generic filter over P. As mentioned in [11, I 1.12], $UG = F$ is a one-one mapping of ω on $\omega_1^{(M)}$. Since the ordinals of N are exactly the ordinals of M the sets constructible in N are exactly the sets constructible in M. Since M satisfies $\mathbf{V} = \mathbf{L}$, all sets of M are contructible in M, and therefore the sets constructible in N, are exactly the members of M. By the construction of N we have, therefore, that $V = L[G]$ holds in N. Since $F \in N$, and since we have, also in N, $G = \{f \in P \mid f \subseteq F\}$, we have $N = M[G] = M[F]$. N is a a model of $\mathbf{ZF}$, therefore N contains a binary relation R on ω which well-orders ω in the order-type $\omega_1^{(M)}$, and a real r which codes R. Since we can recover F, and hence G, from p, we get that $M[r]$ contains F and G, and thus $M[r] = M[G] = N$, and $V = L[r]$ holds in N, which establishes (3) in N.

To see that (23) holds for $\leqq_P$ we proceed as follows. Let π be a permutation of ω, then for $p \in P$ we denote with $\pi_*(p)$ the function $\{\langle \pi(n), p(n)\rangle \mid n \in \mathrm{Dom}(p)\}$ (where, $\mathrm{Dom}(p)$ is the domain of p). π_* is obviously an automorphism of $\langle P, \leqq_P\rangle$. Given $p, p' \in P$, let π be a permutation of ω mapping $\mathrm{Dom}(p)$ on a set disjoint with $\mathrm{Dom}(p')$. Obviously, $\pi_*(p)$ and p' have disjoint domains and are, hence, compatible. Thus (23) holds for $\langle P, \leqq_P\rangle$. Since in N, M is the class of all constructible sets, we get, by Lemma 3, that (2) holds in N.

Let $T \in N$ be an ordinal-definable binary relation which well-orders a set A of real numbers. Every member of A is ordinal-definable, since for every $a \in A$ there is an ordinal α such that a can be defined as "the αth member of the well-ordering T". Since every finite ordinal is obviously definable we get that every $a \in A$ is also *hereditarily*-ordinal-definable in N; thus, by (2), every member of A is constructible in N. In M we have a one-one mapping of the set of all constructible reals on $\omega_1^{(M)}$. This mapping is also in N, and since $\omega_1^{(M)}$ is countable in N, we get that there are only $\aleph_0$ constructible reals in N, hence A is countable in N. Thus we have established that (7) holds in N.

§3. Proof of Theorem 3. Here we start again with a countable transitive model of $\mathbf{ZF} + \mathbf{V} = \mathbf{L}$. For the set P of conditions, we use the conditions used by Cohen [2, p. 130] to add τ subsets to ω, only that we take $\tau = \omega_1^{(M)}$. Later in the paper, we shall need some of the results proved in the present section for a general τ, and, therefore, we shall assume here $\tau = \omega_1^{(M)}$ only where this will be mentioned explicitly. P consists of functions whose domain in a finite subset of $\tau \times \omega$ and which obtain only the values 0 and 1. $\leqq_P$ is the inclusion relation. Let G be an M-generic filter over P, and let $F = \bigcup G$. We have now, in analogy to what we had in the proof of Theorem 2, that F is a function on $\tau \times \omega$ into $\{0,1\}$ and that $F \in N$ and $N = M[G] = M[F]$. By Cohen [2, p. 134, Theorem 4] we have in N, for $\tau = \omega_1^{(M)}$, $2^{\aleph_0} = \aleph_1$. Since M is the class of all sets constructible in N we have in N, $V = L[F]$ and $F \subseteq \omega_1 \times \omega$. This implies that by [7, 2.8 and Theorem 2], $2^{\aleph_\alpha} = \aleph_{\alpha+1}$ holds in N for every $0 < \alpha \in N$. Thus the **GCH** holds in N.

Let $A \in M, A \subseteq \tau$. We denote with P_A the subset of P consisting of all p's whose domain is included in $A \times \omega$, $G_A = \{p \in P_A \mid p \in G\}$, $F_A = \bigcup G_A$. It is easily seen that for a generic G, $F_A = F \mid (A \times \omega)$, where $F \mid B$ denotes the restriction of the function F to the set B.

LEMMA 4. *For every real number $a \in N$, there is a set $A \in M$, $A \subseteq \tau$, which is countable in M, and such that $a \in M[G_A]$.*

Proof. Let $\bar{a}$ be a term of L'' which denotes a. For every statement Φ of L'' the set of minimal conditions p, which force Φ, is countable in M (Cohen [2, p. 134, Lemma 7]). Let Q_n be the set of all minimal conditions p which force $\mathbf{n} \in \bar{a}$. Q_n is a sequence of M of countable sets. Therefore, there is in M a sequence A_n of countable subsets of τ, such that $p \in Q_n \to \mathrm{Dom}(p) \subseteq A_n$. Put $A = \bigcup_{n<\omega} A_n$; A is a countable subset of τ. We claim that for every $n < \omega$,

$$n \in a \leftrightarrow (\exists p \in G_A)(p \Vdash \mathbf{n} \in \bar{a})$$

The right-to-left implication follows from $G_A \subseteq G$. We shall now assume $n \in a$ and prove the right-hand side. If $n \in a$ then $\mathbf{n} \in \bar{a}$ is true in N and hence it is forced by some $p \in G$. Let $q \subseteq p$ be a minimal condition which forces $\mathbf{n} \in \bar{a}$. Since $q \subseteq p \in G$, also $q \in G$. By the definition of Q_n, $q \in Q_n$ and hence $\mathrm{Dom}(q) \subseteq A_n \times \omega \subseteq A \times \omega$, thus $q \in G_A$, which establishes the right-hand side above. We have now $a = \{n \in \omega \mid (\exists p \in G_A)(p \Vdash \mathbf{n} \in \bar{a})\}$ and this is obviously a set of $M[G_A]$.

LEMMA 5. *Let $A \subseteq \tau$; every real number b which is ordinal-definable from G_A is constructible from G_A.*

Proof. P is obviously isomorphic to $P_A \times P_{\tau-A}$, therefore, by Solovay [11, Lemma I 2.3], $M[G] = M[G_A][G_{\tau-A}]$ and $G_{\tau-A}$ is generic over $M[G_A]$. (23) holds for $P_{\tau-A}$, as easily seen. A real number b ordinal-definable from G_A is also hereditarily-ordinal-definable from G_A, and therefore, by Lemma 3 with $M[G_A]$ replacing M, $b \in M[G_A]$. Since $V = L$ holds in M, we get that b is constructible from G_A.

Now we proceed to prove Theorem 3. Our Π^1_2 relation will be "*g is not constructible from f*". That this is indeed a Π^1_2-relation can be seen by the methods used at the end of our §1 (they are due to Silver and Shoenfield). Suppose now, that there is a real-ordinal-definable relation $S \in N$, which uniformizes our relation. Let a be a real number from which S is ordinal-definable. By Lemma 4 and its proof, there is a countable subset $A \in M$ of τ such that a is definable from G_A and $\bar{a}$. $\bar{a} \in M$, hence $\bar{a}$ is constructible, and therefore ordinal-definable, in N. Therefore, a is ordinal-definable from G_A in N, and hence S is ordinal-definable from G_A in N. Since A is a countable set in M, it is easy to construct in N a real number r, which codes G_A (e.g., if h is a constructible one-one mapping of A on ω then we consider the set $\{\{\langle\langle h(\alpha), n\rangle, p(\langle\alpha, n\rangle)\rangle \mid \langle\alpha, n\rangle \in \mathrm{Dom}(p)\} \mid p \in G_A\}$; this is a subset of $R(\omega)$ — where $R(\beta) = \bigcup_{\gamma<\beta} P(R(\gamma))$, and $P(x)$ is the power-set of x — and there are one-one constructible mappings of $R(\omega)$ on ω). Therefore we obtain $M[G_A] = M[r]$ and S is ordinal-definable from r. Let f be the characteristic function of r and let $g \in \omega^\omega$ be defined from S and f as "the g such that $S(f, g)$". Since f and S are ordinal-definable from r, g is also ordinal-definable from r, and hence from G_A. By Lemma 5 g is constructible from G_A. (In fact, Lemma 5 applies to real numbers but we can replace g by a real number coding it). On the other hand, by definition of S, g is not constructible from f, and hence g is not constructible from r, a contradiction.

§ 4. Proof of Theorem 4. Here we start with a countable transitive model M of **ZFCI** + **V** = **L**. We use now the same set P as in Solovay [11, I, §3]. Let Ω be an inaccessible ordinal in M. P consists of all functions p on finite subsets of $\Omega \times \omega$ into Ω such that $p(\alpha, n) < \alpha$. $\leqq_P$ is the inclusion relation. Let G be an M-generic filter over P, and $F = \cup G$. F is a function on $\Omega \times \omega$ into Ω such that for a fixed $\alpha < \Omega$ $F(\alpha, n)$ maps ω onto α. As mentioned in [11], the model $N = M[G]$ thus obtained, is a model of **ZFC** + **GCH**.

If we replace Ω by some ordinal $\lambda < \Omega$ in the definition of P, we get a subset P^λ of P. It is easily seen that we can regard P as $P^\lambda \times Q$, where Q consists of all members of P whose domain is included in $(\Omega - \lambda) \times \omega$. For an M-generic filter G on P let $G^\lambda = G \cap P^\lambda$, $G^* = G \cap Q$. By Solovay [11, I, §3.4, Corollary 1], for every real number $a \in N$ there is an ordinal $\xi < \Omega$ such that $a \in M[G^\xi]$. Therefore a is ordinal-definable from G^ξ. Since $\xi < \Omega$, ξ is countable in N, and hence G^ξ can be coded by a real r in N. Let W be a well-ordering of a set T of real numbers in N which is ordinal-definable from the real number a in N. As we saw in §2, every $b \in T$ is ordinal-definable from a, and is, hence, also ordinal-definable from G^ξ. By exactly the some argument used in the proof of Lemma 5, we get that each $b \in T$ is constructible from G^ξ, and hence from r. By Solovay [11, I, §3.4, Corollary 2] there are in N only $\aleph_0$ reals which are constructible from any given real, therefore the set T is countable in N.

§5. Proof of Theorem 5. Here we start with a countable transitive model M of $\mathbf{ZF} + \mathbf{V} = \mathbf{L}$ in which $\Lambda > 0$ and Λ is not confinal with ω. For the set P of conditions we take the conditions used by Cohen [2, p. 130] to add τ subsets to ω, and we choose $\Lambda^{(M)}$ for τ. These conditions were treated in §3. By [2, p. 134, Th. 4] we have in N $2^{\aleph_0} = \aleph_\Lambda$; by [2, p. 132, Th. 2] (4) holds in N.

LEMMA 6. *The following holds in N: If T is a set of real numbers which is real-ordinal-definable then there is a real number r, such that either $T \subseteq L[r]$, and then, $|T| \leq \aleph_1$, since there are at most $\aleph_1$ reals constructible from a single real, or else $|T| = \aleph_\Lambda$.*

REMARK. Theorem 5 is obviously an immediate consequence of Lemma 6.

Proof of Lemma 6 (beginning). Let $\bar{a}$ be a name in L'' of a real number a of N. Let Q be the set of all minimal conditions which force sentences $\mathbf{n} \in \bar{a}$ for $n \in \omega$, and let $A = \{\alpha \mid (\exists p \in Q)(\exists k \in \omega)\ (\langle \alpha, k\rangle \in \mathrm{Dom}(p))\}$. A is a countable subset of τ in M (see the proof of Lemma 4 in §3). Let us call A the *support* of $\bar{a}$. We shall now prove several lemmas before we shall go on with the proof of Lemma 6.

LEMMA 7. *If $\bar{a}$ is the name in L'' of a real number a of N, and A is the support of $\bar{a}$, then $a \in L[G_A]$ holds in N.*

Proof. It is shown in the proof of Lemma 4 in §3 that $a \in M[G_A]$. Since $M = L$ holds in N this implies our lemma.

LEMMA 8. *Let $\Phi(x)$ be a formula of L' with, no free variables other than x, and which does not contain the symbol G. Let $A \subseteq \tau$ $A \in M$; by $\Phi(G_A)$ we denote a sentence of L' which asserts, in a natural way, that G_A satisfies $\Phi(x)$. If $p \in P$ and $p \Vdash \Phi(G_A)$, then already $p \mid (A \times \omega) \Vdash \Phi(G_A)$.*

Proof. To prove $p \mid (A \times \omega) \Vdash \Phi(G_A)$, we have to show that for every M-generic filter G over P, if G contains $p \mid (A \times \omega)$ then $M[G] \vDash \Phi(G_A)$. By the properties of a generic filter, G contains a condition q such that $\mathrm{Dom}(q) = \mathrm{Dom}(p)$ (since for every $\alpha < \tau$ and $n < \omega$ there is a $k \in \{0,1\}$ such that $\{\langle\langle \alpha, n\rangle, k\rangle\} \in G$ — see, e.g., Solovay [11, I, 3.2] — and q is the union of finitely many such sets $\{\langle\langle \alpha, n\rangle, k\rangle\}$). Since both q and $p \mid (A \times \omega)$

are members of G they are compatible, and since $\mathrm{Dom}(q) = \mathrm{Dom}(p) \supseteq \mathrm{Dom}(p|(A \times \omega))$, we have $q \supseteq p|(A \times \omega)$. Let σ be the automorphism of P in M defined as follows. For $p' \in P$ $\sigma(p')$ is given by $\mathrm{Dom}(\sigma(p')) = \mathrm{Dom}(p')$ and for $\langle\alpha, n\rangle \in \mathrm{Dom}(p')$

$$\sigma(p')(\langle\alpha, n\rangle) = \begin{cases} 1 - p'(\langle\alpha, n\rangle) & \text{if } q(\langle\alpha, n\rangle) \neq p(\langle\alpha, n\rangle) \\ p'(\langle\alpha, n\rangle) & \text{otherwise.} \end{cases}$$

By [11, Lemma 2.1] $\sigma(G)$ is an M-generic filter over P and $M[\sigma(G)] = M[G] = N$. Obviously $\sigma(q) = p$, and since $q \in G$ we have $p \in \sigma(G)$. Since $\sigma(G)$ is M-generic over P, and $p \Vdash \Phi(G_A)$, we have that $\sigma(G)$ "satisfies" $\Phi(G_A)$ in $M[\sigma(G)] = N$, i.e., $\Phi((\sigma(G))_A)$ holds in N. However, since p and q coincide on $A \times \omega$, we have $(\sigma(G))_A = G_A$, and thus $\Phi(G_A)$ holds in N, which is what we wanted to show.

LEMMA 9. *Let $A, B \subseteq \tau$, $A, B \in M$, and let $x \subseteq \omega$, $x \in M[G_A] \cap M[G_B]$; then $x \in M[G_{A \cap B}]$ (see §3 for the notation).*

Proof. Since $x \in M[G_A]$ there is a term u of L'' which denotes x in $M[G_A]$. u itself belongs to M. Let $\Phi_1(y, z)$ be a formula of L' which asserts in a natural way that y is a member of the set of $M[z]$ denoted by the term u. (u occurs in $\Phi_1(y, z)$ as a constant $\mathbf{u}$ since L' has individual constants for the members of M.) $\Phi_1(y, z)$ does not contain the symbol G. By our choice of $\Phi_1(y, z)$ we have

(24) for every $n < \omega$, $N \models \Phi_1(n, G_A)$ if and only if $n \in x$.

Similarly, since $x \in M[G_B]$ there is a formula Φ_2 of L_2 such that $N \models (\forall n \in \omega)((N \models \Phi_2(n, G_B)) \leftrightarrow n \in x)$ and hence $N \models (\forall n \in \omega)(\Phi_1(n, G_A) \leftrightarrow \Phi_2(n, G_B))$. ($G_A$ is not a symbol of L' and thus $\Phi_1(n, G_A)$, e.g., obviously abbreviates an appropriate formula of L'). Therefore, there is a $p \in G$ such that

(25) $$p \Vdash (\forall n \in \omega)[\Phi_1(n, G_A) \leftrightarrow \Phi_2(n, G_B)].$$

We claim now that

(26) $$x = \{n \in \omega \,|\, (\exists q \in G_{A \cap B})(q \cup p \Vdash G_1(\mathbf{n}, G_A))\}.$$

In order to prove (26), we assume that n is in the right-hand side of (26). Since $q \in G_{A \cap B} \subseteq G$ and $p \in G$ we have $q \cup p \in G$ and then $\Phi_1(\mathbf{n}, G_A)$ is true in N, and therefore, by (24), $n \in x$. Going in the other direction, assume $n \in x$, then for some $p' \in G$ $p' \Vdash \Phi_1(\mathbf{n}, G_A)$. Since $p, p' \in G$ p and p' are compatible and we can assume, without loss of generality, that $p' \supseteq p$, and thus we can represent p' as $p \cup q''$. We have now $p \cup q'' \Vdash \Phi_1(\mathbf{n}, G_A)$. Let $q' = q''|(A \times \omega)$; by Lemma 8 we have $p \cup q' \Vdash \Phi_1(\mathbf{n}, G_A)$. The set of sentences forced by a condition is closed under implication in **ZF**; therefore we get from $p \cup q' \Vdash \Phi_1(\mathbf{n}, G_A)$ and (25) that $p \cup q' \Vdash \Phi_2)\mathbf{n}, G_B)$. Let $q = q' \mid (B \times \omega)$; by Lemma 8 applied to $p \cup q' \Vdash \Phi_2(\mathbf{n}, G_B)$ we get $p \cup q \Vdash \Phi_2(\mathbf{n}, G_B)$. This together with (25) implies $p \cup q \Vdash \Phi_1(\mathbf{n}, G_A)$. $q = q' \mid (B \times \omega) = (q'' \mid (A \times \omega)) \mid (B \times \omega) = q'' \mid ((A \cap B) \times \omega)$. Since also $q \subseteq p' \in G$ we have $q \in G_{A \cap B}$. Thus we have shown that n belongs also to the right-hand side of (26).

By the absoluteness of the relation of forcing we see that (26) defines the set x in $M[G_A]$, hence $x \in M[G_A]$.

Proof of Lemma 6 (end). T is in N a real-ordinal-definable set of real numbers. Therefore there is a formula $\Psi(x, y)$ of L', and a real number $a \in N$, such that

$$T = \{b \subseteq \omega \mid b \in N \wedge N \Vdash \Psi(a, b)\}.$$

(The ordinals in the definition of T occur in $\Psi(a, b)$ as individual constants of L'). Let A be the support of $\bar{a}$. Since A is countable in M, G_A can be coded in N by a real number r. If we do not have $T \subseteq L[r]$ in N then let $b \in T$ be a real number not constructible from r, and hence from G_A, in N. Let $\bar{b}$ be a name for b in L'', and let B be the support of $\bar{b}$; $\bar{b}, B \in M$. Since we have in N $b \subseteq \omega \wedge b \notin L[G_A]$, there is a condition $p \in G$, such that

$$(27) \qquad p \Vdash \bar{b} \subseteq \omega \wedge \bar{b} \notin L[G_A] \wedge \Psi(\bar{a}, \bar{b}).$$

Since B is countable in M and τ is uncountable in M there is a family ρ_α, $\alpha < \tau$, of permutations of τ such that: for $\alpha < \tau$ and $\gamma \in A$ $\rho_\alpha(\gamma) = \gamma$; and for $\alpha, \beta < \tau$, if $\alpha \neq \beta$ then $\rho_{\bar{\beta}}^{\,-1}(B - A) \cap \rho_\alpha^{-1}(B - A) = 0$, and hence $\rho_{\bar{\alpha}}^{\,-1}(B) \cap \rho_{\bar{\beta}}^{\,-1}(B) \subseteq A$. For every $\alpha < \tau$ and for every finite subset s of $\tau \times \omega$ we define a permutation $\sigma_{\alpha,s}$ of P as follows. For every $q \in P$, $\mathrm{Dom}(\sigma_{\alpha,s}(q)) = \rho_\alpha(\mathrm{Dom}(q))$, where $\rho_\alpha(\mathrm{Dom}(q))$ is taken to be

$$\{\langle \rho_\alpha(\gamma), n \rangle \mid \langle \gamma, n \rangle \in \mathrm{Dom}(q)\}; \text{ and for } \langle \gamma, n \rangle \in \mathrm{Dom}(q)$$

$$(\sigma_{\alpha,s}(q))(\rho_\alpha(\gamma), n) = \begin{cases} q(\gamma, n) & \text{if } \langle \gamma, n \rangle \notin s \\ 1 - q(\gamma, n) & \text{if } \langle \gamma, n \rangle \in s. \end{cases}$$

Let $\mathrm{val}(G, u)$ be the function of N which gives the value of the term u of L'' when we assign to the constant G the subset G of P. $\mathrm{val}(G, n)$ is defined for every $G \subseteq P$ and for every term u of L''. Let $F(\alpha, s) = \mathrm{val}(\sigma_{\alpha,s}(G), \bar{b})$. By Lemma 7, $F(\alpha, s) \in L[(\sigma_{\alpha,s}(G))_B]$ in N. Since $(\sigma_{\alpha,s}(G))_B$ is obtained from $G_{\rho_{\bar{\alpha}}^{-1}(B)}$ by the permutation $\sigma_{\alpha,s}$ which is clearly a member of M we have

$$(28) \qquad L[(\sigma_{\alpha,s}(G))_B] = L[G_{\rho_{\bar{\alpha}}^{-1}(B)}] \quad \text{in } N.$$

and therefore

$$F(\alpha, s) \in L[G_{\rho_\alpha^{-1}(B)}]. \tag{29}$$

Since G is generic and ρ_α^{-1} (Dom(p)) is a finite subset of $\tau \times \omega$, there is a unique $p_\alpha \in G$ such that $\mathrm{Dom}(p_\alpha) = \rho_\alpha^{-1}(\mathrm{Dom}(p))$. We have, thus, $\rho_\alpha(\mathrm{Dom}(p_\alpha)) = \mathrm{Dom}(p)$, and hence also

$$\mathrm{Dom}(\sigma_{\alpha,s}(p_\alpha)) = \mathrm{Dom}(p). \tag{30}$$

Let $s_\alpha = \{\langle \gamma, n\rangle \in \mathrm{Dom}(p_\alpha) \mid p_\alpha(\gamma, n) = 1 - p(\rho_\alpha(\gamma), n)\}$. p_α and s_α are sequences of N (and not necessarily of M). Throughout the rest of the proof of Lemma 6 our aim will be to show that $F(\alpha, s_\alpha)$ is a one-one mapping of τ into T in N. Having established that we shall know that T has at least τ members in N. By our choice of τ, $\tau = \aleph_\Lambda^{(M)}$; by the absoluteness of Λ, and by the absoluteness of the notion of cardinal ([2, p. 132]), we have $\tau = \aleph_\Lambda^{(N)}$ and our lemma is proved.

By our definition of $\sigma_{\alpha,s}$ we have

$$(\sigma_{\alpha,s_\alpha}(p_\alpha))(\rho_\alpha(\gamma), n) = \begin{cases} p_\alpha(\gamma, n) & \text{if } \langle\gamma, n\rangle \notin s_\alpha \\ 1 - p_\alpha(\gamma, n) & \text{if } \langle\gamma, n\rangle \in s_\alpha \end{cases}$$

$$= \begin{cases} p_\alpha(\gamma, n) & \text{if } p_\alpha(\gamma, n) = p(\rho_\alpha(\gamma), n) \\ 1 - p_\alpha(\gamma, n) & \text{if } p_\alpha(\gamma, n) = 1 - p(\rho_\alpha(\gamma), n) \end{cases} = p(\rho_\alpha(\gamma), n)$$

Since $\sigma_{\alpha,s_\alpha}(p_\alpha)$ and p have the same domains, by (30), and, by what we have just shown, they obtain the same values for the same arguments, we have $\sigma_{\alpha,s_\alpha}(p_\alpha) = p$. Since $\sigma_{\alpha,s}$ is a permutation of P and belongs to M, $\sigma_{\alpha,s}(G)$ is M-generic over P and $M[\sigma_{\alpha,s}(G)] = M[G] = N$, by Solovay[11, I, Lemma 2.1]. Since $p_\alpha \in G$ and $\sigma_{\alpha,s_\alpha}(p_\alpha) = p$ we have $p \in \sigma_{\alpha,s_\alpha}(G)$. As consequence of the genericity of $\sigma_{\alpha,s_\alpha}(G)$, of $M[\sigma_{\alpha,s_\alpha}(G)] = N$, of $p \in \sigma_{\alpha,s_\alpha}(G)$, of (27) and of $F(\alpha, s_\alpha) = \mathrm{val}(\sigma_{\alpha,s_\alpha}(G), \bar{b})$, we get in N

$$F(\alpha, s_\alpha) \subseteq \omega \wedge F(\alpha, s_\alpha) \notin L[(\sigma_{\alpha,s_\alpha}(G))_A] \wedge \Psi(\mathrm{val}(\sigma_{\alpha,s_\alpha}(G), \bar{a}), F(\alpha, s_\alpha)). \tag{31}$$

By (28), $L[(\sigma_{\alpha,s}(G))_A] = L[G_{\rho_\alpha^{-1}(A)}]$; but since ρ_α is the identity on A we have $L[\sigma_{\alpha,s}(G))_A] = L[G_A]$. Therefore, by (31), $F(\alpha, s_\alpha) \notin L[G_A]$. We claim now that for $\alpha, \beta < \tau$, if $\alpha \neq \beta$ then $F(\alpha, s_\alpha) \neq F(\beta, s_\beta)$. To see this suppose that $F(\alpha, s_\alpha) = F(\beta, s_\beta)$. By (29) we have $F(\alpha, s_\alpha) \in L[G_{\rho_\alpha^{-1}(B)}] \wedge L[G_{\rho_\beta^{-1}(B)}]$ in N.

By Lemma 9 $F(\alpha, s_\alpha) \in L[G_{\rho_\alpha^{-1}(B) \cap \rho_\beta^{-1}(B)}]$ (since $M = L$ in N). By our definition of ρ_α we have $\rho_\alpha^{-1}(B) \cap \rho_\beta^{-1}(B) \subseteq A$ (as we have mentioned right after the definition of ρ_α), therefore $F(\alpha, s_\alpha) \in L(G_A)$. ($\rho_\alpha^{-1}(B) \cap \rho_\beta^{-1}(B) \in M$ and hence $G_{\rho_\alpha^{-1}(B) \cap \rho_\beta^{-1}(B)} \in L[G_A]$.) However, we have shown above that $F(\alpha, s_\alpha) \notin L(G_A)$, thus our assumption $F(\alpha, s_\alpha) = F(\beta, s_\beta)$ leads to a contradiction, and therefore $F(\alpha, s_\alpha)$ is a one-one function of α.

Our proof will therefore be finished if we show, that for every $\alpha < \tau$ $F(\alpha, s_\alpha) \in T$; and to show this we have to show that $\Psi(a, F(\alpha, s_\alpha))$ holds in N. By (31) we have in N, $\Psi(\mathrm{val}(\sigma_{\alpha,s_\alpha}(G), \bar{a}), \bar{b})$, hence it will be enough if we prove here $\mathrm{val}(\sigma_{\alpha,s_\alpha}(G), \bar{a}) = \mathrm{val}(G, \bar{a}) = a$. We shall prove that $(\sigma_{\alpha,s_\alpha}(G))_A = G_\alpha$; since A is the support of $\bar{a}$ we have

$$\begin{aligned} \mathrm{val}(G, \bar{a}) &= \{n \in \omega \,|\, (\exists p \in G_A)(p \Vdash \mathbf{n} \in \bar{a})\} = \\ &= \{n \in \omega \,|\, (\exists p \in (\sigma_{\alpha,s_\alpha}(G))_A)(p \Vdash \mathbf{n} \in \bar{a})\} \\ &= \mathrm{val}(\sigma_{\alpha,s_\alpha}(G), \bar{a}), \end{aligned}$$

which is what we had to show. All that is left to prove now is $(\sigma_{\alpha,s_\alpha}(G))_A = G_A$. We shall prove it by showing that $\sigma_{\alpha,s_\alpha}(q) = q$ for every $q \in G_A$. By our definition of $\sigma_{\alpha,s}$, $\mathrm{Dom}(\sigma_{\alpha,s_\alpha}(q)) = \rho_\alpha(\mathrm{Dom}(q))$. Since $q \in G_A$, $\mathrm{Dom}(q) \subseteq A \times \omega$ and since ρ_α is the identity on A, we get $\mathrm{Dom}(\sigma_{\alpha,s_\alpha}(q)) = \rho_\alpha(\mathrm{Dom}(q)) = \mathrm{Dom}(q)$. If $\langle \gamma, n \rangle \in \mathrm{Dom}(q)$ then, by $q \in G_A$, $\gamma \in A$ and hence $\rho_\alpha(\gamma) = \gamma$. Therefore, by the definition of $\sigma_{\alpha,s}(q)$,

$$(\sigma_{\alpha,s}(q))(\gamma, n) = \begin{cases} q(\gamma, n) & \text{if } \langle \gamma, n \rangle \notin s \\ 1 - q(\gamma, n) & \text{if } \langle \gamma, n \rangle \in s. \end{cases}$$

Once we show that $s_\alpha \cap \mathrm{Dom}(q) \subseteq s_\alpha \cap A \times \omega = 0$, we will have established that $(\sigma_{\alpha,s_\alpha}(q))(\gamma, n) = q(\gamma, n)$, for every $\langle \gamma, n \rangle \in \mathrm{Dom}(q)$; and this, together with $\mathrm{Dom}(\sigma_{\alpha, s_\alpha}(q)) = \mathrm{Dom}(q)$, which we have already shown, implies $\sigma_{\alpha,s_\alpha}(q) = q$, which is what we had to show. To prove $s_\alpha \cap A \times \omega = 0$, let $\gamma \in A$; if $\langle \gamma, n \rangle \notin \mathrm{Dom}(p_\alpha)$ then, by definition of s_α, $\langle \gamma, n \rangle \notin s_\alpha$; if $\langle \gamma, n \rangle \in \mathrm{Dom}(p_\alpha)$, then $p_\alpha(\gamma, n) = p(\alpha, n)$, since both p_α and p are in G, and are hence compatible, and therefore, again by the definition of s_α, $\langle \gamma, n \rangle \notin s_\alpha$, which concludes our proof.

REFERENCES

[1] J. W. Addison, *Some consequences of the axiom of constructibility.* Fundamenta Mathematicae **46** (1959), 337–357.

[2] P. J. Cohen, *Set Theory and the Continuum Hypothesis.* New York, 1966.

[3] S. Feferman and A. Levy, *Independence results in set theory by Cohen's method II (abstract)*, Notices of the Amer. Math. Soc. **10** (1963).

[4] A. Hajnal, *On a consistency theorem connected with the generalized continuum hypothesis.* Acta Mathematica Academiae Scientiarum Hungaricae **12** (1961), 321–376.

[5] A. Levy, *Definability in axiomatic set theory I.* Logic, Methodology and Philosophy of Science, proceedings of the 1964 International Congress (Y. Bar-Hillel Ed.), Amsterdam, 1965, pp. 127–151.

[6] A. Levy, *Independence results in set theory by Cohen's method I, III, IV (abstracts).* Notices of the Amer. Math. Soc. **10** (1963), 592–593.

[7] A. Levy, *A generalization of Gödel's notion of constructibility*, Journal of Symbolic Logic **25** (1960), 147–155.

[8] J. Myhill and D. Scott, *Ordinal Definability.* Proceedings of the 1967 International Symposium on set theory held in Los Angeles, to appear.

[9] J. R. Shoenfield, *On the independence of the axiom of constructibility.* Amer. J. of Math. **81** (1959), 536–540.

[10] J. R. Shoenfield, *Mathematical Logic.* Reading, Mass., 1967.

[11] R. M. Solovay, *A model of set theory in which every set of reals is Lebesgue measurable*, to appear.

[12] E. Specker, *Zur Axiomatik der Mengenlehre (Fundierungs- unda Auswahlaxiom).* Zeitschrift für mathematische Logik and Grundlagen der Mathematik **3** (1957), 173–210.